Power Plant Construction Management

POWER PLANT
CONSTRUCTION
MANAGEMENT

A Survival Guide
2ND EDITION

Peter G. Hessler

Copyright© 2015 by
PennWell Corporation
1421 South Sheridan Road
Tulsa, Oklahoma 74112-6600 USA

800.752.9764
+1.918.831.9421
sales@pennwell.com
www.pennwellbooks.com
www.pennwell.com

Marketing Manager: Sarah DeVos
National Account Executive: Barbara McGee Coons

Director: Mary McGee
Managing Editor: Stephen Hill
Production Manager: Sheila Brock
Production Editor: Tony Quinn
Book Designer: Susan E. Ormston
Cover Designer: Karla Womack

Cover photo courtesy of KOG Transport, Inc., USA.

Library of Congress Cataloging-in-Publication Data

Hessler, Peter G.
  Power plant construction management : a survival guide / Peter G. Hessler. -- Second edition.
      pages cm
  Includes bibliographical references and index.
  ISBN 978-1-59370-337-0
  1. Power-plants--Design and construction--Management. 2. Electric power-plants--Design and construction--Management. I. Title.
  TJ164.H47 2014
  690'.54--dc23
                                        2014031030

Contents

Preface

This book, the second edition of the original *Power Plant Construction Management: A Survival Guide* has been written to update and upgrade its predecessor. Almost a decade has passed since the first edition. A lot has transpired, and sometimes re-transpired, during those years. As discussed in more detail in the introduction, since the first edition was published the construction industry in general, and the power plant construction industry in particular, have had several ups and several downs. The construction industry went through major shifts in the availability and cost of labor, supervision, equipment, and materials. It has started to embrace advanced technology, tools, and processes made available with the exponential advance of computers (especially their downsizing) and computing power, intelligent cell phones, and the Internet. And during this same period, the power plant industry has seen shifts to and from coal-fired generation, gas-fired generation, solar, wind, nuclear, and other renewables. In other words, the only constant during this intervening decade has been change.

As was the case for this book's first edition, this second edition has also been written to provide economic guidance and support to those involved in the management of power plant construction activities, whether these activities are the building of a new plant or the rebuilding, repowering, or modification of existing components. This book is not intended to be a technical "how to" manual on performing construction work, on selecting the mechanical tools and equipment for executing the construction activities, or on directing the day-to-day activities of the work. Instead, this edition, just like its predecessor, is intended to provide a stimulus to perform these tasks in a cost effective manner—to think outside of the box—to think about the financial stakeholders of the project/s at hand. In other words, this book will not help the reader to select the right crane for a particular lifting task, but it will offer the reader support in deciding whether using a crane in the first place is the most economic thing to do.

Although the book title and its supporting examples are specifically power plant oriented, the fundamental theories and practices discussed within are applicable to any construction endeavor, from power plant construction to road building, from refinery construction to chemical process plant expansions, and for commercial and municipal projects. The underlying premise is that the construction activities being undertaken are being performed for the ultimate benefit of the owner/stakeholder, and this owner/stakeholder expects a return for investing in this endeavor.

Still today, too many projects, within and outside of the power plant industry, do not meet the expectations of their stakeholders from an economic perspective. Some of the reasons for this can be traced back to events that preceded the start of site activities, and for this reason this second edition includes information and provides examples of what to do, and not to do, early in the project cycle. It goes into the initial decision-making process of whether a project should even be undertaken from a construction risk point of view. It covers the development and understanding of construction specifications. And it drives deep into the budgeting and estimating phase of preparing for a construction project. The first edition only covered this material from a 30,000-foot level.

As pointed out by many of the contributors to this book, maintaining awareness of the finances of the site construction activities is extremely important, but being able to predict—and correspondingly impact—the outcome, early during the project, is even more crucial. As one contributor, John Long, who is now retired from Constellation Energy, put it, "Surprises are not acceptable." In other words, the outcome may be inevitable, but it is crucial to be able to predict this early on; the owner/stakeholder must be afforded the opportunity to mitigate.

The first edition was written to provide support for several different circles. First, it was a treatise on managing the economics of power plant construction, intended to be useful for the site superintendent to get from today to tomorrow. Second, it was intended for the site general manager to direct his staff in performing their duties in a coordinated and focused manner. But third, it was also written to provide the management of all site personnel a primer on what to expect from their charges and how to offer them ideas and support. The first edition book provides detailed formats for accomplishing many of the tasks of performing construction management, and it offers examples of how to use these formats in various settings. This second edition does the same, even using some of the same language and examples, but it also enhances the information in the first edition by extending the circles to include much more of the pre-site work. Both are designed to be useful in teaching environments. In fact, the first edition has been used as the foundation for many one- and two-day construction management workshops ever since it hit the bookshelves in 2005. It is intended that this new revision will be the foundation for many more workshops, both in academia and in industrial settings.

As a final note, this second edition does not specifically address working outside of North America. The first edition did. That is not to say that the information in this second edition is not applicable to working elsewhere in the world. It is. But when planning and managing construction projects beyond our borders, a lot more is required. There are cultural impacts. There are different legal implications. There are skill level challenges. There are distances, terrain, and seasonal hurdles to overcome. The list goes on and on. Therefore, if one is seriously considering a project outside of North America, the first edition of this book is still a must-read.

As the author, I hope you, the reader, benefit from the contents of the book you are now holding. I hope you will share the contents with your peers. As a former coworker Gary "Red" Wilcoxon once said about the first edition: "Many can bid and win, but few can track and execute. If you find the problems fast you have time to react, even if the work is fast-paced. The tools are in this book—Read it and use them." I look forward to meeting some of you as time marches forward, and I look forward to hearing from you—to hear how some parts of this book made your jobs successful as well as where things did not work out so well. Please let me know at PGHessler@ConstrBiz.com.

Acknowledgments

Just as was the case for the first edition, the second edition of this book would not be possible without the support and input from many people. A topic as broad as managing the construction activities of power plant work is made up of many parts, and no one person can be an expert on all of them. Seldom is the expert on putting together the project process also an expert on safety or quality control. Seldom is the expert on contract structuring and commercial framing also an expert on site staff management and managing the site activities. So the preparation of this book, just like its first edition predecessor, depended heavily on many persons who graciously donated their time, providing expert material and reviewing various pages of the author's manuscript.

In many instances throughout this book, recognition has been given directly to the contributors of expert quotes, graphs, and figures. This is particularly true for the many contributions by Mark Bridgers, a business associate and personal friend of the author. But there are also many instances where support was provided on a broader basis, and it is this support that the author also wishes to identify and recognize.

First, the basic premise of the first edition, and now the second edition, could not have materialized without the active participation of the publisher's staff; their guidance and support made both books possible. Second, the author's concept of focusing on the cost elements of the construction process was strongly influenced by many individuals with whom he has worked collectively for many, many years—individuals such as John Long, Gary (Red) Wilcoxon, and Ron Blodgett, among others. Their willingness to take time from their respective, very busy schedules to provide input, review manuscript content and offer suggestions was invaluable.

Then, there were other individuals who provided support for very specific subject areas such as risk management, suretyship, the quality process, and safety; these subject areas would have floundered without their guidance. Risk management is a cornerstone of both books and is referenced directly and indirectly many times; this would have been very difficult without the collaboration of two of the author's closest business associates, Bob Tichacek and Mark Bridgers. In the first edition of this book, the author depended on the support of Surety Information Office's Marla McIntyre to explain the importance to the site staff of understanding how the surety and insurance process works and that often decisions made (or not made) by this staff can have economic consequences directly related to this subject. That subject was reinforced again in this second edition.

The author also gratefully acknowledges the support of a former business associate, Jim Pillow, for the guidance and clarifications provided during the development of the chapter on controlling quality. Since the cost impact of quality is often an imprecise measurement, this subject would also have appeared imprecise without this help. Another former associate, Dick Peterson, was instrumental in helping the author clarify the impact various safety practices have on the cost of the site work. Too often, safety is only respected due to regulatory pressures but with this guidance, the author was able to present safety as a cost control device as well.

Special recognition is also due to now retired Jim DeGraff, former supervisor of outage and project management at Constellation Energy. Jim spent countless hours doing a peer review of the initial manuscript. Drawing on his many years of power plant project management expertise, Jim offered invaluable suggestions that helped streamline the subject matter and therefore made the final manuscript a much more cohesive product.

Finally, the author very gratefully thanks his wife, Teresa, for her whole-hearted support during the long period of preparing the manuscript. The many days and nights of "being absent" from the normal routine of life were often frustrations that she endured with love and compassion. Without her support, this book would not have met any of the deadlines that seemed to arrive all too often.

There were many, many others whom the author would also like to thank, but in the interest of time and space, they will remain unnamed, although not unremembered. Thank you one and all for your help and support!

Introduction

When the first edition of this book was written nearly a decade ago, the power plant construction industry had just undergone a major transformation. It had gone from long-term, large-scale, coal-fired power plant build-outs to fast-track, smaller-sized, gas-fired, combined-cycle plant builds, mostly due to low gas prices at the time.

But after the first edition was published, the price of gas went through the roof, and these gas-fired projects came to a screeching halt. The industry actually reversed itself and started looking at major coal-fired projects again. For example, in 2007, the U.S. Department of Energy's National Energy Technology Laboratory (NETL) published a study suggesting that about 100 GW (gigawatts) of additional power capacity would be needed by the year 2020, and almost 50% of that was to be coal-fired.[1] These coal-fired plants alone amounted to nearly 150 new projects. The increase in the economic activity of the country and the increase in population growth at that time was creating an insatiable demand for more power, a demand that seemed to be opening up one of the busiest times ever in the power plant construction business. That was then.

But, wow, what a change just one year later, let alone two, three, and more! That pending construction boom had everyone extremely busy getting ready for it. Even I, while working on this second edition book, was forced to postpone its completion while working with owners and contractors to move forward for this unprecedented, impending boom. But then, suddenly, the U.S. power plant construction world turned upside-down again. The political landscape went into a state of flux, no longer providing a clear view of where power plant owners could prudently take risks. Climate control legislation was unpredictable, leaving many owners in the dark as to what pollution control technologies would be required. They could have spent hundreds of millions of dollars, only to find out that the newly installed pollution control equipment did not remove the pollutants that might be legislated out in the future. So owners stopped planning for large, new coal-fired plants.

Then, starting in 2008, power plant project cancellations became endemic. Costs for the construction of these planned or in-progress plants, especially the coal-fired plants, had started escalating at an out-of-control pace. According to the February 14, 2008, edition of the electronic version of *Power Engineering*, "the cost of new power plant construction in North America increased 27 percent in 12 months, and 19 percent in the most recent six months."[2] This led to a change in the business model for constructing power plants. Construction firms were no longer willing to commit to fixed-price contracts. Instead, they were attempting to shift the risks of these higher material and construction prices to the plant owners. And on top of this, the "Great Recession" hit, so regardless of the cost of building power plants, money was suddenly just not available. The result is that of the 150 coal-fired power plant projects reported in the 2007 NETL study, a lot fewer were or will be built. Instead, in their place will be more gas-fired plants, along with various renewable power generation projects.

According to a follow-on NETL study published in 2012, the additional power capacity build-out between now and 2020 will have a fuel mix that has shifted dramatically from essentially a 50/50 coal-fired/gas-fired combination to a 30/70 ratio, with gas-fired generation expected to be the plant of choice.[3] There are two specific reasons for this. The first is the dramatic increase of available shale gas, heretofore inaccessible, through the use of a technology called fracking; this has driven the price of gas so low that the economics dictate gas over coal. The second reason is the high cost of additional equipment that will be required for new coal-fired plants to capture CO_2 in the future, thereby eliminating them from consideration for many future projects.

Had this book gone to press as originally scheduled, it would have been based on obsolete facts, out-of-date data, and conditions that no one ever anticipated. Now, the reader can take comfort in the knowledge of these recent, turbulent conditions. That is not to say that other, unforeseen issues will not arise in future years. Certainly, they will, so one must always be ready to adapt, that is, have a Plan B. But the need for prudent construction management will always be there, whether for new-build coal, gas, or renewables; retrofits; transmission lines for electricity and fuel; or just general maintenance outages. As Tony Licata, retired vice president at Babcock Power Environmental, Inc., put it:

Since the first edition of this book was published, our industry has installed 100,000 MW of FGDs [flue gas desulfurization systems] and 140,000 of SCRs [selective catalytic reduction systems] in coal-fired power plants. Almost all of this work was done on brown field projects in retrofit applications. The application of these technologies in most cases was on extremely difficult construction sites in operating power plants which required a new approach to construction project management. The insights to project management in that first edition book have been a very useful tool in our approach to these projects. Managing projects in this new scenario also has forced a change to the contractual terms and conditions over this period, requiring updating our approach to the market from firm price to alliance type projects and then back to firm price projects.

In other words, the only constant is change, and therefore project and construction managers must be able to adapt to these changing conditions. That is the purpose of this second edition, to discuss how to maintain a prudent construction management process, even as the underlying forces are shifting.

Today, It's Still All About the Money

No longer do the rules of the 1960s, '70s, and '80s apply. Today's set of rules governing the power generation industry are the rules of economics. The power generation industry today has owners that place a greater emphasis on the return on their investment than owners did in the past. The industry has changed forever. Even where there is still protection through regulation, owners, including shareholders, want to see a return on their investment *now*. The former protected guarantee of blue-chip stock returns from the major utilities has gone the way of the industrial giants of yesteryear. Today, to stay in business, a company must provide a return to its shareholders that exceeds what they could earn elsewhere. In essence, any business today is only as viable as the edge its return on investment has over other options those investors may care to explore; most power companies are now viewed as expendable by their shareholders. The power plant business of today is about much more than just generating megawatt hours, it is also about generating profits for the investor.

Planning a power plant project and planning the construction activities of a power plant project in today's environment require a financial focus. Equipment, technology, and operational skills have improved since the last building boom. But now there is a shortage of skilled labor and skilled supervision to embark on major building programs. Between 1990 and 2010, a generation of power plant construction skills was lost. This will require that a different approach to the planning and execution of these new projects is used.

According to various sources within the industrial construction community, the average age of a certified pipefitter construction welder is approximately 42. The new apprentices entering the workforce are young and inexperienced, and it takes 10 years or so before they are at their peak performance. Therefore, working smarter is necessary. Preplanning in the early stages of a job has become a necessity. Preplanning just-in-time deliveries is important to facilitate smaller footprints and laydown needs, as well as for controlling the cost of inventory control and storage fees. Not only is an emphasis on safety morally correct, but it is also a major factor in the bottom-line labor costs. The list goes on and on.

Using a Managed Process

As utilities start planning, they do so with an eye on the return on their investment. They select the technology, the fuel, and the plant site with this in mind. The same applies when they select the participants and plan the process. Today, more than ever, the preplanning of the construction phase of a power project will impact the total costs of that project, whether it is building a new plant or rehabilitating an old one. Although the construction phase occurs at the end of the project process, it is really the tail that wags the dog. The project concept may start in operations, it may start in maintenance, or it may start in engineering. But then it moves into budgeting, and from there to project management and the construction preplanning phase. And when it gets to the field, the cost of change, the cost of inefficiency, and the cost of cancellation can be devastating. Therefore, the path to success is to link all of these phases by preplanning the process so it can be managed toward a successful conclusion.

To manage a power plant construction project, one must look at three basic elements:

- Pursuing and ensuring a manageable contract (or contractor)
- Structuring the contract correctly
- Managing the contract to a successful conclusion

The first edition of this book only addressed the latter of these three elements. This second edition addresses all three. It uses examples that are more in line with tomorrow's expected power mix, such as a heavier emphasis on gas-fired generation. The data have been updated to reflect current-day numbers and the trend toward more technology. Preplanning for a construction project requires a dedicated and managed process; this book covers how to achieve that. The need for adequate resources to preplan is addressed. Contingency planning is covered from a view of factually determining how much, if any, contingency should really be included.

The budgeting, bidding, and estimating phases are introduced. The concept of managing this stage of the project, as if it were the site work itself, is discussed. Developing and writing, as well as reading and understanding, the specifications come next. There are many different thoughts on how this should be done, and these are compared. Designing the work scope and reviewing its logic is explored, and developing the resultant schedule and constraints is explained with examples of what can happen if not properly done.

Time is spent discussing how to develop price requests, how to develop pricing based on these requests, and how to structure the contract so pricing can be changed, if necessary, as the project goes forward. This is then translated into payment requirements. Taxes are addressed, and the pros and cons of penalties and bonuses are shown, with examples. Also, there is a detailed discussion of the importance of cash flow.

The estimating process is looked at from both the owner's perspective and the contractor's viewpoint. The importance of a site visit is highlighted with checklists. Labor, tools, and supervision are reviewed from a perspective of current-day costs and availability, and then there is some discussion about how to extrapolate these costs into the future. Additionally, the cost savings of proper quality control and safety planning are put into perspective, and finally, the methods of arriving at the total budget, or price offering, are described.

Next, planning the delivery structure of the site works is addressed. Things change during the overall project timeline, and even though the delivery structure may have been envisioned one way during the preplanning stages, it needs to be reviewed in light of actual conditions that exist as the work begins. As an example, resources such as labor, supervision, and tools and equipment may no longer be as available as was once thought. This is addressed, along with training.

Commercial terms and conditions are explored. Examples are given to show what can happen, or not, if the people in the field do not understand these requirements. This then leads into a chapter on risk management. Preparing for thorough risk management at the job site is addressed by discussing the risk management stool and how it is supported by its three legs: the insurance coverage, the claims management process, and the contract wording. This book explores all three.

Next, there is a chapter on setting up the site itself, followed by a chapter on managing quality and safety from a financial perspective. In fact, the central theme of this second edition is identical to the successful original edition, managing the financials of building a power plant (or any other heavy industrial project, for that matter). In the next three chapters tools are provided to assist in setting up and managing the finances of the site works. There are checklists, guidelines, photos, and examples that can be used immediately for managing the work activities and reporting results, up and down the line. By using these tools, the decision making process can be greatly enhanced.

Finally, there is a chapter on how to use current-day technology to make the work of managing the process at the job site much easier. Since this technology is constantly evolving, suggestions are offered as to how to stay on top of it.

It is the author's hope that the reader will find the topics herein to be of use in his or her own daily practice. As stated in the original edition of this book:

> I've taken all the body of knowledge that I've come across over the years, combined that with the observations of how things seem to really work out on the street, and the result of that is what I've come up with in this book.[4]

Read on and enjoy.

References

1 *Tracking New Coal-Fired Power Plants.* Washington, DC: National Energy Technology Laboratory, U.S. Department of Energy, May 2007.

2 "Power Plant Construction Costs Rise 27 Percent," *Power Engineering.* Electronic version, February 14, 2008.

3 *Tracking New Coal-Fired Power Plants.* Washington, DC: National Energy Technology Laboratory, U.S. Department of Energy, January 2012.

4 Ott, Richard. *Creating Demand.* Homewood, IL: Business One Irwin, 1992.

Preplanning to Planning 1
(with content from Mark Bridgers)

As discussed in the Introduction, preplanning cannot be overemphasized. Many instances of project failure can be traced directly to the lack of project preplanning. When an owner ponders the pros and cons of embarking on a project, or when a contractor looks for projects to bid on, some fundamental criteria should be in place, both technical and economic. If the project does not meet these criteria, the owner needs to rethink the plans and the contractor needs to be willing to say "no" to the opportunity. This is the start of the preplanning process.

However, the preplanning process needs a preplanning team. The team members must be knowledgeable about the purpose of the project, and they need to understand the impact the project can have on the organization. In the course of their work, they will develop an understanding of the project's risks and rewards, and they will assist management in deciding whether this project should be undertaken, and if so, why and how (for both the owner and the contractor).

But, where do the resources come from to do all of this? Funding may not even be available yet. So how do you pay for a preplanning team? Where do preplanning support groups charge their time? What are the earlier mentioned fundamental criteria? Who decides? How? When?

Once the preplanning team is in place, it should consider looking at past projects—those that were successful and those that were not. Part of the preplanning process is learning from the past. Preplanning should be a process where the project specifics are investigated in enough detail to form overall execution strategies. Local conditions, resources, and impacts on the community should be reviewed. Permitting should be addressed. (See chapter 4 for more on permitting).

Then, the preplanning team should move on to considering the project delivery system. Should there be a formal alliance, a clear subtier contracting arrangement, or what some in the industry now call *coopetition*? Or does

1

one go back to the old days of design-bid-build, construction management at risk, or some other variation? In other words, how should the project delivery be structured? And what does one do when it is not possible to determine this up-front?

But this just scratches the surface. The budgeting process needs to be addressed early on. Where will the funds come from during the construction phase? How shall they be quantified? And what about contingencies? Do they form part of the budget, or are they to be treated differently? How to answer these questions will be clearer after reading this chapter.

The Preplanning Team

To answer the above questions, one of the first steps is putting together a preplanning team. But how does this preplanning team get formed? Where do the people come from? How are they funded? Who supports them? Who is their champion? What will be their charter? All are very important questions.

But since the core question to be answered, upon completion of this team's work, is "Do we pursue this project?" a process for the team (owner or contractor) to follow needs to be put in place. Most likely, the process will look something like this:

- Clarify and document the preplanning goals.
- Suggest preplanning action plans.
- Design a preplanning control process (measure and manage).
- Recommend whether the opportunity should be pursued or not pursued.

An illustration of this can be seen as follows: A construction contractor is looking for more business. The marketing or sales group identifies an opportunity with an architectural engineering (AE) firm shopping for contractors to build a 1,000-MW combined-cycle gas-fired power plant in a sparsely populated area where the summers can be severe.

Suppose this happens to be a contractor with many years of experience, including boom and bust cycles and good and bad jobs. If so, this would imply that there have been occasions to learn from the past. Prudence therefore would suggest that before jumping at the chance to go after this project, a structured evaluation should be followed, that is, the preplanning phase.

This means that a preplanning team should be formed. In this case, it will consist of three core team members: someone from sales who is familiar with the AE and plant owner; a representative from the field staff, familiar with the practical challenges the construction manager will face if the job is awarded; and someone who will be responsible for pulling together the proposal and the bid if the project is pursued. These three preplanning team members will be charged with the responsibility to investigate this project, evaluate its value to the contractor's organization, and make recommendations for a go/no-go decision.

Their charter could be very informal, but the questions these team members would be expected to answer as they evaluate the project's impact on the corporation would be very specific. They would set specific goals. For example, one goal could be to evaluate the potential for profit in keeping with corporate budgets. Another could be to determine if the execution of the project would enhance the corporate image within the industry. A further goal could be to identify major risks and analyze the potential for mitigating them. There should be a goal to investigate future opportunities with this client, if it was a new client. And finally, there should be a goal to compare the needs of the project with the availability of company resources.

The action plans of the preplanning team should include a review of the specifications prepared by the AE and a summarization of the major requirements. They should develop a rudimentary list of work packages such as the following:

- Earthworks scope
- Civil and structural steel scope
- Mechanical work
- Electrical requirements
- Instrumentation and controls
- Start-up support

They should be expected to identify the amount of effort required to perform this work, and whether or not it will fit within the normal business expertise of the organization.

Next, they should be expected to itemize any significant risks that might arise, such as schedule constraints and the potential for incurring liquidated damages. They also should be responsible for investigating the difficulties of working in the area, considering labor sources, support infrastructure, and climate.

In parallel, one team member's action plans could be talking with the AE and end user or owner of the plant regarding their plans for other contractors, labor resources and management, site services, and all of the other items that are required to coordinate a project of this magnitude. One of the major considerations at this stage of the preplanning would be to determine if the effort required to bid on a job like this is worth the expense. Maybe the proposal and bidding efforts could be used more effectively for other opportunities. It is not uncommon for a contractor to spend hundreds of thousands of dollars chasing a large project.

To keep the team on track, a control process would then be set up. A schedule of the action plans would be developed and accepted by each team member. It would be designed so that if the final decision is to go forward with the bidding process, there would be adequate time for this to be accomplished.

As this process goes forward, there is usually a mentor within the organization who ensures that the team stays on track. This mentor, or champion, would remove obstacles that might be in the way of the team obtaining temporary resources to evaluate various scenarios to determine project risks. This champion would be someone appointed at a high level of the organization, someone who was clear on the goals for meeting the increased growth of the business. In fact, often this person would be the one that makes the final go/no-go decision, based on the recommendations of the preplanning team.

Learning from the Past: The Lessons Learned Process

Next, the preplanning team should look to the past. It is strange how we like to reinvent the wheel, over and over, when it comes to managing a project, and especially when it is the construction phase that's being managed. Using the excuse of time constraints, we seldom look back. We take a project, we assign some staff, and we tell them to march forward.

Yet the managers of really successful jobs are those who do not just march forward. They come to a stop, they push back. They first do some research on earlier projects and then compare the basics between the old and the new. They look up the members from the earlier projects. They prepare an outline of their own challenges and then assemble a group of representatives of the present and the past to review the lessons learned from the previous project.

In preparation for this, the successful project managers review the requirements of their project of the day. Then they prepare a status of where they are versus where they want to go. Maybe the specifications have not yet been written. If not, there may be an opportunity to include or exclude items that were problems in the past. If the specs have already been prepared, maybe there will still be time to issue addenda to accomplish the same goal. Maybe the work divisions have not yet been assigned. Depending on local conditions, job-site conditions, commercial considerations, and political ramifications, there may be an opportunity to reduce the chance of future problems. By polling the lessons learned group attendees, many fresh ideas can be generated.

What to review

For example, work scope reallocation could be in order. Let's say that the installation of the underground utilities was usually assigned to a stand-alone contractor. But after hearing of difficulties in scheduling on some previous projects, there may be a decision to reallocate the installation to the earthworks contractor, even if this contractor subsequently subcontracts the work; the risk would thereby be shifted onto the contractor most closely tied to the schedule of access (and success), the one most able to manage this specific risk.

Local conditions often dictate how a job should be set up and how it should be run. First and foremost is weather. Is the job site located in a bitterly cold climate? Is it in the dry heat of the desert? Or is it in an area subject to lots of rain? Climatological data can help in determining the impacts weather may have on the job. If schedule is important, a review of previous jobs, under similar conditions, could reveal innovative ways to weather the proverbial storm. Suppose the job is in an area known for heavy rains. If this were a grassroots project, heavy rains could easily impact the earthwork phase. Reviewing previous jobs in similar situations might reveal that installing a special, but temporary drainage system at the outset of the work would greatly enhance dewatering. This could then reduce the time required for dewatering and possibly save days on the schedule allocated for the earthworks.

Maybe the job-site conditions are not yet well known during the preplanning stage. Reviews of similar projects by the lessons learned group can highlight which site conditions are critical. For example, if the job requires moving a lot of large components from the off-loading point or from a preassembly area, traffic routing and traffic patterns would be critical. Imagine a large heat recovery steam generator (HRSG) being transported to site from a barge unloading dock, as in figure 1–1. When it comes time to transport this, whole roads may be blocked for hours at a time. What would this do to the other work that also needed the same roads for access?

Fig. 1–1. Transporting this HRSG blocks road and access for hours. (Courtesy of KOG Transport, Inc., USA)

What about resources? Many construction jobs fail because of the lack of preplanning for resources. The most important of these is the labor. Labor is the single most costly, complex, and yet most important part of almost every power plant construction job. For example, if some of the people from a previous job have any insight into the source and competency of labor for the current project, their stories of what went right and wrong could be invaluable.

There is another factor that will work against the project or construction manager when trying to get the site works up and running: younger workers' lack of job skills. Many enter construction jobs without any formal training beyond high school. Given that the industry has become increasingly complex, with the proliferation of advanced technology and tools employed on job sites, highly skilled workers are vital. Unfortunately, the young people typically considering craft work often lack the necessary math, communication, and technical skills required—leading to increased training costs and decreased productivity.[1] Again, preplanning for a way to overcome the lack of these skills is necessary very early during the project inception stage.

After labor, both the supervision and the tools and equipment needed also will have a decided impact on the bottom line. Therefore, it behooves the team to look at these items in some detail. If this is a job with a lot of complicated lifting requirements, then a combination of ideas from both groups in the meeting will be beneficial. Sometimes, a revision of plans, such as installing a temporary trolley system or using mobile cranes on the roof of the structure in lieu of using expensive off-road crawler cranes, can save time and money. Other times, smart planning, such as using two smaller hydraulic cranes to transport a steam drum, can save money and headaches (fig. 1–2). And it can lead to additional flexibility in project execution.

Then there are commercial issues that can impact a construction project incessantly. For example, if the owner's specifications require the contractors to pay large—$100,000 plus—per-day liquidated damages and the schedule is tight, maybe there will be an experience from a previous job that can aid in mitigating this. Envision an issue similar to the above HRSG road access issue, but without an alternative for avoiding job slowdowns during the movement of large components, for example, preassembled ductwork. Even if a review of previous projects does not produce a solution, the presence of the reviewing participants can often generate fresh ideas. In this case, someone may recognize that the job is close to an area where there are heavy-duty helicopters to be used for the lifts. Or maybe there is the possibility of laying railroad tracks for some of these activities, thereby not needing roads for access.

Fig. 1–2. Drum positioning with hydraulic cranes

Last but not least, some projects get caught up in local, and not so local, politics. It could be that the job is not on the list of favorite projects in the area. Nuclear comes to mind. Wind turbine blades' noise and bird impacts are others. Resultant protesters can have a negative impact on labor productivity and site access and create a host of additional problems.

As a mitigating strategy, arranging local town hall meetings to explain the upside values of the project can help. If economic conditions warrant, pointing out the increase in employment for the duration of the job can sometimes offset protests. Making special efforts to develop a good rapport with the local law enforcement personnel would also be of value. Thinking ahead is important. Once again, a review of earlier projects with similar issues can assist in developing solutions to these kinds of issues.

How to review it

There are various ways to proceed through a lessons learned process, and it is important to select the one most likely to ferret out the maximum options for the project at hand. The project participants can simply huddle around a table and talk. Or, they can use a preplanned agenda. They can use a list of questions drawn up by both sides prior to the meeting. But one of the most effective ways is to use a facilitator, with a basic outline of concerns for success. Then, as the discussions commence, the facilitator can (a) keep the discussions on track and (b) maximize participation by all involved. For example, the facilitator could have the participants write ideas on sticky notes and post them on the wall under predetermined headings.

Let's say that part of the currently planned project is to convert multiple, existing control rooms to one consolidated control room. A natural concern would be about the sufficiency of the cabling system. In other words, can existing cable trays and conduit be used? If new trays and conduit are to be run, how large should they be, and what size and how many cables will be required? Therefore, one of the predetermined headings could be "cable trays and conduit." Participants could then be asked for their ideas on the problems envisioned and potential solutions. Some of these ideas could be:

- Wait until engineering is complete before purchasing materials and starting work.
- Purchase a variety of trays and conduit to fit all potential needs.
- Invite the design engineers to the meeting.
- Plan to route and install oversize trays and conduit and load them with spare cables.

With these ideas, and maybe more, the participants could jointly determine the effect of the various ideas upon schedule, cost, and personnel. If time was of more concern than cost, the last idea, oversize trays with spare cables, might be the preferable course. But, if time was not of the essence, the first or third options might be the best. Because of their previous experience, personnel who worked on earlier projects can help to analyze the solutions more clearly.

Planning for the Project Delivery Structure

At the same time that the lessons learned process is proceeding, preplanning for the project delivery structure should be underway. In fact, the project delivery structure decision should not be finalized until it has been vetted as the most complimentary structure for managing the issues discussed during the lessons learned sessions.

In the distant past, the design-bid-build delivery mode was used almost exclusively. Utilities would develop (design) the specifications and then solicit bids for the work. After evaluation of the bids, the lowest price contractor would be awarded the job, and the utility's construction group would manage the contractors. This was a costly and lengthy approach, but it gave the utility total control of the process, something many felt was required since the management mandate was to focus on reliable power generation. Profitability was already guaranteed through the regulatory process.

However, as utility regulatory groups started focusing on the *price* of power in addition to *reliability*, utility managers began to look for more cost-effective ways to build power plants by better managing their construction projects. This led directly to contracting with AE firms for the development of the specifications, the solicitation of the contractor bids, and many times, for the management of the field construction activities as well. Since many of the AE firms were often involved in multiple power projects at any one time, their personnel developed economies of scale that resulted in lower design and construction management costs than if the utility performed these functions itself.

Near the end of the 1980s and in the early 1990s, changes in regulations governing the generation and sale of electricity encouraged more and more cost control. This led to the creation of a new kind of power generator, the independent power producer (IPP). These IPPs were focused solely on selling electricity at a profit, quickly, with their selling prices reflecting the supply and demand of power. Therefore, they were very focused on controlling costs and minimizing the construction schedule, especially when building new plants.

This led to a major shift in the basic type of plant delivered—a shift from large, steam-driven turbines to smaller, gas-fired turbines that could handle rapid load shifts at higher machine efficiencies. These plants could be delivered, installed, and commissioned much sooner than their steam-driven predecessors.

But to accomplish this, these IPPs needed to find alternate ways of managing their build-out projects. So they turned to delivery methods such as cost-plus not to exceed maximum price, construction manager as contractor (CMc), or fast-tracking. Each was intended to speed up the project cycle from inception of the project to generation of power. No longer was the old design-bid-build method suitable.

However, times continued to change. Shortly after the start of this century, the rise in the price of fuel for these gas-fired turbines forced power generators to once again look at other power generation systems. As referenced in the Introduction to this book, suddenly plans were being made once again to build lots of new coal-fired power plants, with huge price tags attached. To contain these costs, some utilities were forced to contract the work under firm price, or cost-not-to-exceed scenarios. This sometimes became necessary just so they could guarantee a fixed price to be used for electricity rate increase requests (to recover the cost of building the power plant). Once again, the method of delivering projects was changing because some of the previous project delivery structures were not the right models anymore.

Today and beyond

But then, just a few years later, gas prices dropped precipitously. Owners were back to ordering more gas turbines again, along with huge increases in planning for wind, solar, and nuclear power generation. Coal-fired generation was suddenly out of vogue once again. It is this kind of seesawing that the planners of the next generation of power plants now find themselves facing.

So what do they do? If it is a grassroots, expansive project, it is of paramount importance to be able to approximate a total project budget, which means planning on a particular delivery structure at the outset. If it is a smaller, less costly maintenance and repair project, the level of budget may not be as important (the budget has probably been predetermined through the annual budgeting cycle), and therefore the determination of the project delivery structure also may be less important.

But if it is a grassroots, expansive project, the financiers of such an undertaking will want to know how much money they are expected to provide. They will want to know how long they will have to wait for a return on their investment. They will want to know what that return will be—and they will want to compare this to a known alternative that may be time dependent. Therefore, even if the project delivery structure has not yet been determined, at this point of the preplanning process, an assumption, and possibly commitment as well, may have to be made to move forward.

The Project Delivery Structure

Once the decision has been made to go forward, the next step will be to determine how. *How*, in this instance, is basically deciding what kind of project delivery structure would make the most sense for the project at hand. The vocabulary around project delivery structure is very loose, and there are frequent misunderstandings of what is meant. The Construction Management Association of America (CMAA), the Design Build Institute of America (DBIA), the Associated General Contractors of America (AGC), and the Construction Users Roundtable (CURT), among others, have attempted to standardize the vocabulary usage to moderate success. The most successful projects have, by design or happenstance, settled on a "highly aligned" project delivery approach. By contrast, the least successful projects have used a "poorly aligned" approach.

The concept of alignment is critical. Outlined below is a simple approach that can help utility and power generation asset owners achieve this alignment in their project delivery approach. In addition, the underlying matrices that help owners apply this approach bring a logical approach to the selection of a project delivery system that will withstand scrutiny.

There are three core questions that the owner must answer: (1) How will I manage the design and construction effort? (2) How will I build the asset? (3) How will I contract for services?

The first question about how the effort will be managed describes the degree of hands-on or hands-off approach of the owner. As was described previously, in the distant past it was not uncommon for the utility to exercise total control of the design and construction process, or what in figure 1–3 is described as multiprime and requires a very hands-on approach by the owner, including contracting with all of the trades and service providers directly and taking responsibility for coordinating their work at the site. On the opposite end of the spectrum in figure 1–3 is what is referred to as program outsource. This is where the owner hires a firm to execute the capital construction efforts with the owner serving in a hands-off capacity while the asset is designed and constructed.

Fig. 1–3. How will you manage?

The second question (fig. 1–4) speaks to the quality and level of construction drawings while the project is under construction. The traditional approach is design-bid-build, in which the owner prepares or has prepared a complete set of construction drawings that in theory reflect all conditions and accurately described how the facility will be constructed. Again, in theory, when an owner undertakes an effort in this fashion, there should be no change orders or adjustments in scope necessary. The opposite end of the spectrum is described as design-build, in which the construction drawings are prepared in parallel with construction activity in order to avoid having to prepare a complete set of construction drawings before pricing and construction can begin. Design-bid-build requires a trade-off of time spent to achieve accuracy and detail in the construction drawings, while design-build sacrifices some accuracy and detail in order to start construction sooner. In addition, in the design-bid-build approach, the owner typically contracts for design services and construction services separately, with separate firms, and at different times, whereas design-build typically results in a contract for both design and construction services with the same firm at the same time.

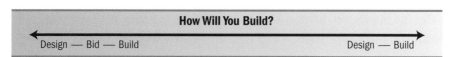

Fig. 1–4. How will you build?

The third question (fig. 1–5) speaks to how the services will be contracted, and given the typical ratio of 1 or 2 to 10 between owner spending on design and later construction services, construction carries more weight in getting the contracting approach aligned. With this said, the fact that design services occur as the first set of activities and can have such a significant impact on ultimate construction cost outcome, they must be considered. The spectrum of choices for the owner range from hard-bid to negotiated. Hard-bid refers to a situation where the only consideration for selecting a design or construction service provider is the price of its services. This approach is typically only applied in the public sector where municipalities either cannot (by law) or do not differentiate among service providers. Prequalifying or select bid is much more typical outside of the public sector, where some type of validation of capabilities takes place before even considering a service provider. A negotiated approach typically balances the price of services with other characteristics the owner has defined as critical in the selection of a service provider. These might include safety performance, past experience, financial capacity, willingness to enter into a collaborative relationship, and so forth.

Fig. 1–5. How will you contract?

Alignment—and why it matters

Alignment in the selection of a project delivery system results in incentives and expectations being more clearly communicated and implemented. In answering the three questions above, owners should seek responses that tend to be on the same side of the spectrum in order to achieve alignment. In a public or municipal setting, it may be impossible for owners to achieve this type of alignment, given the legislative restrictions under which they have to operate. In this case, they may have to use other techniques to bring a higher degree of alignment after procurement has taken place. The best example is the use of partnering concepts.

In a private setting, owners typically do not face external restrictions or requirements that dictate project delivery system characteristics. Given that there are many different project delivery structures available, selecting an appropriate approach is challenging. However, all of them must address the three core questions posed previously. Most of the time, the decision of which type of structure has a direct relation to the entity's appetite for risk.

Some years ago, many owners divested themselves of their in-house project engineers and project management teams. To compensate for this, they contracted with turnkey groups that accepted the responsibility for most of the project risk. It was not unusual in this type of setting to use a design-build approach with a general contractor with a hard- or select-bid.

Unfortunately, this approach yields a project delivery system that is not highly aligned and, in many instances, results in an escalation of construction costs. Many engineering, procurement, and construction (EPC) contractors shifted away from this risk acceptance, forcing some of these owners to re-engage in the management arena. But the way they have done this, in many instances, has not been to return to the past. Instead, they have turned to collaborative and integrated delivery structures that can be more flexible and result in more aligned project delivery systems with shared visions and risks. Some of these more collaborative and integrated project delivery structures are described below.

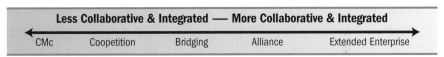

Fig. 1–6. Collaboration

Construction manager as constructor (CMc)

The evolving project delivery structure has been embraced outside of the power plant construction industry for some time. Now, it is emerging as one of the project delivery structures in building power plants. Although similar to the traditional design-bid-build method, it differs in that the CMc is brought in early in the design phase of the project. The intent is to get input from the constructibility perspective, which should then save costs once the field work gets underway. Additionally, the CMc can help to speed up the overall project delivery process.

But the most important result of using a CMc project delivery structure is to be able to lock in the construction cost. By being intimately involved in the up-front planning and design work, the CMc will be able to help in the development of a firm price for the field construction work early in the project, usually before the site work gets very far along. However, as with any project delivery structure, the CMc approach used by one owner may be different from the way it is used by another.

For example, just because the CMc is involved near the beginning of the project planning phase, they or the owner may still be uncomfortable about accepting a simple firm price commitment for the site works. For one thing, the CMc frequently is brought on board without having gone through a formalized bidding process. Therefore, the construction price actually may be structured as an incentive-based or shared over/under (target price) contract. Also, it is here where building information modeling (BIM) and integrated project delivery (IPD) work most effectively.

One misconception about the CMc process, however, is that it works automatically. It does not. It will only work when the participants in the project care more about the project than their corporate goals. It really means that their corporate goals must be aligned to the success of the project, first and foremost.

Coopetition

Coopetition (cooperative competition) came to the attention of the author during a McGraw-Hill Construction's ENR [*Engineering News-Record*] Top Firm Leaders Forum held in New York City some years ago. It was during the presentation by a team composed of individuals from Bovis Lend Lease, the Port Authority of New York & New Jersey, and Fluor Corporation that this term was explained as an alternate project delivery structure. This team was using coopetition for some of the World Trade Center rebuilding, but suggested that it should be considered by other project groups as well. Their basic premise was that coopetition breeds fresh and refined ideas. In this mode, the best practices of each partner are used in the project, and each partner therefore has the opportunity to learn from the others. Since often the partners are also competitors on other projects, there does exist the question: How far into your particular inner company workings are you willing to let the other person look? According to Bob Prieto, senior vice president of Fluor Corporation,

> It's all a matter of balance. While databases and proprietary tools can be used in a controlled way on cooperative projects, unfettered access to the underlying software and databases will likely not happen. That is where the real intellectual property is at the end of the day. However, the sharing of processes, procedures and lessons learned is much easier to accomplish, and in some ways results in a more permanent transfer of best practices. These can include: sharing approaches to risk management, where a structured, comprehensive approach may be adopted but underlying databases, checklists and contingency factors are only shared at a summary level:

- adopting management processes for safety oversight and executive inspection of project safety practices where partners can adopt best practices without having access to the proprietary underpinnings

- agreement on basic techniques for executing construction activities in the field relating to how forms are attached, on-site materials managed or rigging utilized

Coopetition on projects benefits owners, but also the project participants who can now reassess their own systems, procedures and tools with the benefit of the perspective of other knowledge-able industry players.

Bridging

Going one step further, there is another project delivery structure that also should be considered: bridging. According to the website of the Bridging Institute of America (BIA), "bridging is a construction project delivery method designed to reduce the owner's risks and costs in quality construction." It structures the roles of architects, consulting engineers, general contractors, and subcontractors, so that the owner gets a firm construction price much earlier in the design process and does not surrender the control of the design, engineering, or construction.

Per the BIA website, if properly executed, the construction price for the owner can be more dependable than the tried-and-true design-bid-build method. This is, in part, due to a major reduction in change orders. What is different from other project delivery methods is that the owner or the program manager selects a contractor and also brings on board an owner's representative to provide third-party support on behalf of the owner. The contractor is then charged with designing and building the project. The owner's representative is there to act as a "bridge" between the owner and contractor.

BIA says that following their process can reap savings of 4%–5% in contract prices and provides the owner with a fixed construction price in about half the normal time and at half the normal design cost. For more detailed information on this project delivery method, go to the BIA website at www.bridginginstitute.org.

Alliances

Alliances are making a comeback. Although sometimes misconstrued as a means of outsourcing, a properly structured alliance between owners and contractors can be much, much more. For example, it can be used to share risk in proportion to ability, oftentimes much more effectively than with a straightforward subcontract. The key to the success of an alliance arrangement is to both design an effective governance structure and implement a set of metrics that clearly show the value of continuing, or not continuing, the arrangement. As John Long, former president of Constellation Power Generation, tells it:

> As I look at alliances, one of the major critical success factors is that both companies have to have a relatively balanced opportunity to benefit (profit) from the alliance. For the Supplier, the benefits are more business (obviously) and a better understanding of how the Owner makes decisions which adds value with other customers. For the Owner, it is tangible cost reductions, productivity/process improvements and/or reductions in planned and forced outages.
>
> Another critical success factor is that both companies need to be open and honest with each other and avoid the "throw the Spec over the wall and get a quote in response" mentality. Transparency of costs and benefits is key.

Another viewpoint comes from an interview between Brian Schimmoller, previously managing editor of *Power Engineering* magazine and executives from the Tennessee Valley Authority (TVA) and Day & Zimmermann NPS (DZNPS) in late 2005, titled "Alliance Reliance." According to Michael McMahon, president of DZNPS,

> Proven, measurable results and lower total cost of ownership (TCO) are the keys. Many plant owners are realizing that both firm-price and blanket time-and-material (T&M) maintenance contracts do not promote teamwork between owner and contractor to drive costs out of projects and produce optimum results. Lump sum work is often inflated to cover contingencies and risk, and administration of such contracts can lead to adversarial relationships that are not in the best overall interest of the project and a long-term relationship between owner and contractor. A blanket T&M contract can be effective, but if a long-term relationship is

not established based on trust and commonly shared goals, the owner may need to closely monitor the agreement to ensure high productivity. No incentives exist for the contractor to work fewer hours; therefore, the contractor attempts to earn more fee by working as many hours as possible.[2]

One of the main issues in setting up a suitable project delivery structure is to find the most cost-effective structure, and this often requires flexibility. An alliance can provide for this. Since the construction contract, by its very nature, is a risk-allocation tool, an alliance structure usually allows each party to assume only those risks that it is in a position to control. It promotes a "no-blame" culture reinforced by financial incentives to achieve specific project goals. Additionally, an often overlooked advantage of an alliance relationship is the cost-saving potential of utilizing a single insurance coverage for the total entity, removing or reducing the need for each party to provide individual coverages that often overlap. And finally, the successful alliance structures could include a "no dispute/no litigation" understanding. This, in itself, would serve to focus everyone's attention away from contract language and onto project goals. In the end, a project delivery structure that is based on an alliance arrangement is a structure that has gone from risk shifting to risk sharing, often a much more cost-effective solution to delivering a project.

Extended enterprise

In the process of trying to mitigate construction risk, many constructors and owners, even if in an alliance, still have a difficult time figuring out how to achieve breakthrough improvement once the typical and easily identified changes are implemented. One way that is advocated by some in the construction industry is to go a step beyond a formal alliance and build a more robust extended enterprise. This approach demands greater integration among the multiple parties involved: typically the owner, designer, contractor, union representatives, and some vendors. It is also typical that a mentor will play some role. This mentor's responsibility is to help the parties forge a partnership that results in the unification of goals and efforts.

This mentor assists in helping the parties come together, entwining their management, personnel, and decision making. The mentor is tasked with helping the senior executives at the respective participating firms to establish a set of common goals and efforts and, more importantly, get the mid- and low-level staff to buy into the process. This is seldom an easy task since paradigm shifts are inherently difficult within any group, let alone groups that historically have been wary of one another. Because of this, an extended enterprise process takes longer to become effective. Its maximum value is not normally realized until one to two years later, but it can yield greater than 15%–20% improvement in cost performance.[3] For this reason, it is best used on long-term projects or over a program of construction that contains multiple projects.

One example of such a process, as described in Bridgers et al., is to help owners in existing relationships re-energize themselves to achieve breakthrough performance or overcome some major obstacle.[4] In this setting, it is not unusual for the parties to have already accomplished some performance improvement and to have run out of what they consider implementable ideas, believing further change and improvement are unattainable. Improvement of unit- and total-cost performance approaching 40% is possible—but only when an owner organization brings the following to the table:

- Strategic intent: A clear definition and communication of the utility's strategic intent
- Total cost of ownership: A deep, detailed understanding of the total cost of ownership
- Internal process gaps: A clear understanding of gaps in the utility's organization that could be filled by a strategic sourcing relationship
- Potential provider assessment: A clear understanding of the potential providers of "gap-filling" services for the utility
- Senior level motivation: Motivation at the executive level of the utility's organization to make the necessary changes to implement a sourcing strategy

Once these characteristics exist, the path to build an extended enterprise that yields transformational change follows 10 steps:

1. Establish a common vision.
2. Build trust.

3. State the transition.

4. Link people, ideas, and processes.

5. Align communications.

6. Engage key players.

7. Stimulate ideas.

8. Start the rapid process innovation.

9. Build on short-term success.

10. Track process innovation.

With these 10 steps, an effective extended enterprise can be formed that will yield transformational performance improvement.

Not any one of the above methods of delivering a power plant construction project is better than the others all of the time. An analysis of owner culture and project characteristics and risks needs to be performed before a final decision can be made as to which delivery system is the best for a particular project, and it is beyond the scope of this book to delve into this. However, some method, either one of the above or some variation, will have to be chosen before the planning process (as opposed to preplanning process) for the actual construction work can begin.

Budgeting: Funding the Work

Funding the preplanning efforts

Even before considering how to budget or fund a project, thought must be given to the funding of the initial preplanning efforts. As discussed earlier, up to three full-time people could be involved in this effort. In many organizations, especially if they are the owners, it is frequently a scramble to figure out how to pay for this preplanning effort. Since the decision to move forward may still be a long way off, or may never be made, project-specific funding is seldom available at this early stage. Yet these preplanning efforts must be paid for, and prudent owners carry such funds in a developmental category, similar to a reserve fund. Not doing so can put undue burdens on the team members assigned to the preplanning efforts.

For a contractor, the situation is somewhat different. Most contractors, especially those who work on several projects at a time, have full-time marketing and sales teams that frequently fulfill the role of preplanners. They are part of the contractor's permanent staff and therefore are fully funded.

Funding after preplanning

Once the decision has been made to go forward with a project, an important task to be addressed is funding. Contractors do not spend a lot of time determining how a construction project will be budgeted. They usually enter into a contract that is essentially cash neutral, requiring minimal cash outlay before revenues start flowing. If that is not the case, then the cost of the borrowing required to finance the work will be built into the price of the project. Owners, however, often face a different situation. They will be the ones paying for the contractors to do the work, while not generating revenues until the work is complete, at times three or four years down the road. Thus, it is important for both the owner and the contractor to understand what funding the work is all about.

How does an owner go about funding a long-term construction project? Let's stay with the earlier example of building a 1,000-MW power plant. That's a lot of work. Lots of man-hours. Lots of supervision time. Many weeks of expensive equipment, from the earthmoving plant to the heavy lift cranes, and from specialized welding and alignment tools to all of the trucks, forklifts, and mobile hydraulic cranes required. The total cost for a gas-fired combined-cycle plant of this size can easily reach close to $1 billion, and for a coal-fired plant, the price tag could be nearly three times as much. Of this amount, the field installation portion may run in the neighborhood of 40%, or up to $400 million over a two- to three-year period. That's a lot of financing that has to be obtained from somewhere.

As opposed to the financing obtained for the land and equipment purchases, there are no assets covering the spending during the construction phase. This is what makes financing the construction phase different. The funds used to obtain the hard assets—the land and equipment—are backed by those assets. The funds needed to actually construct the plant have no backing, or recourse in the event of a default of the project. In other words, the funds used for the construction phase are usually obtained through "nonrecourse" loans. These loans are usually long-term, with the intent of payback being through the revenues generated from the operation of the plant once it is completed. The risk is that (1) the completion of the plant is delayed, thereby delaying the operation of the plant and its subsequent revenue stream, or (2) the plant is not completed, never runs, and therefore never generates a revenue stream. Either way, the lender of the funds for the construction phase of the project has to consider these possibilities, and therefore will charge a carrying fee, or interest rate, commensurate with the risk.

In order to get an attractive interest rate, the owners or borrowers of the construction funds need to demonstrate that they can actually complete the plant on time, and that the plant can generate adequate revenues to repay these funds. That requires being able to convince the lenders that the construction and commissioning teams have the ability to do just that. It also requires showing that there is a market for the eventual power that will be generated, normally done through evidence of a long-term power purchase agreement (PPA). Discussion of PPAs is outside of the scope of this book. For the purposes of this book, we will assume that a market exists, that the economic conditions are conducive to obtaining loans, and that contracts for selling the power are in place. Now, not only do the owners have to demonstrate their capabilities, but their site team and the contractors that will do the actual construction work also have to be able to prove theirs.

Back to the 1,000-MW gas-fired plant. Usually, the owner has money in place to get the engineering done. There is money to purchase the equipment and buy the site, if not already owned. But then there will be the need to finance the construction phase. Assuming a 30-month construction schedule, and assuming a $400 million construction budget, the owner will be looking at various options. If the owner is a public entity, say, a municipality, part of these funds may come from bonds floated for 20 or 30 years. Part of the money may come from federal loans, sometimes tied to using specific technologies, and sometimes through economic assistance grants. Since recent history has shown that the costs of construction can fluctuate significantly, public entities sometimes look for help from the private sector.

Private sector support can take many forms. Some of these may dictate the project delivery structures discussed earlier. Or, to ensure a return of their money, the lenders may require a BOOT (build, own, operate, transfer) arrangement, in which the lenders control the revenue generation stream for a predetermined period after completion of the plant—a period long enough for them to recover their loans and the costs of carrying them. Private funds, whether for publicly owned plants or privately and/or publicly traded owners, can come from as many sources as there are people to come up with funding ideas. Banks used to be a primary source. Today, there are also many venture capitalists willing to invest in these plants, often for the hard property as well as for the soft construction phase. Due to the large sums involved, quite often there is a team of lenders or multiple sources of funds. But regardless of who the lenders are, public or private, they want to recover their cash outlays and the interest that was charged. Referring again to the hypothetical 1,000-MW plant, a $400 million construction loan over a 30-month period will generate in the neighborhood of an additional $60 million in interest, given a rate of about 5.5%. (For a coal-fired power plant, with all of its pollution control equipment, anticipate three to four times more principal and interest.) Besides the cost of the construction itself, the interest charges of the construction loan, which are time dependent, also add up to large numbers, and they are a direct reflection of the speed of completion of the site works. In other words, there may be times when it is more cost-effective to work additional workers or hours to shorten the overall project timeline—or maybe even offer a bonus for early completion, just to keep the construction loan interest manageable!

Contingency Plans

The final piece of the preplanning-to-planning phase should always be a backup plan–plan B. To paraphrase Robert Burns in his poem "To a Mouse," the best laid plans of mice and men often go awry. In other words, *things just will not happen as they were planned*. Many things can change. There can be delays to the project, either at its start or during its execution. Other projects may suddenly surface that will create a demand for the same resources at the same time. A review of lessons learned from previous projects can identify still more issues that might arise. Therefore, a contingency plan should always be part of the up-front plan. Usually, the contingency is in the form of money at the preplanning phase, although it can be converted to time once more details become apparent. Let's look at some examples in the next sections.

Project delays

All too often, the project may not start when originally planned. This may be due to any number of causes from difficulty in obtaining permits to problems with bonding issues to changes in funding, and many more. Management personnel that were originally scheduled for the project may now no longer be available because they may be on another project. Labor can be in short supply when the project finally does start due to the same reasons, and this also happens with some of the resources such as heavy equipment. Some type of contingency plan must be considered, even if not yet priced.

First, if obtaining permits becomes more difficult than originally envisioned, one backup plan, plan B, could be to consider engaging a third party to work through the permitting process. If issuing bonds becomes politically sensitive, consideration may be given to taking the cause directly to the people, independent of the governmental bodies that are in office. This can be done in many ways, from using the newspaper, radio, or television media to holding town hall meetings.

Next, if the supervisory personnel suddenly were not available, a good preplanning team would work up a contingency plan that, after determining the most mission-critical positions, would keep these people gainfully employed until the job materialized. Let's look at some numbers: Assuming six months of delay, four people at a cost to the company of $100 per hour would cost about $400,000. Compared to a project with liquidated damages of $500,000 per calendar day, this is less than one day of overrun, a smart contingency plan for ensuring that the right people are on the job when they are needed!

But what about the labor? Here, there are a lot more people to deal with. The complexity of this issue is impacted by the planned source of the labor—is it union, non-union, subcontracted, in-house, or something else? The key is to have a plan to stay in close communication with the source or supplier of these resources. If it is ever evident that there *may* be a delay, this needs to be discussed and contingency plans developed. If the delay is for only a few weeks, one possibility could be to encourage the craftsmen to take vacation. If the delay is of longer duration, there are other options such as paying a bonus to those who make themselves available when needed, importing craftsmen from outside the area and paying them a per diem for food and lodging or changing the planned source of labor from direct hiring to subcontracting, or vice versa.

Finally, if the original plan included the use of equipment such as large excavating machines or heavy lift cranes, their availability must also be reaffirmed. In these cases, it is sometimes possible to pay a "supplier guarantee" fee in return for assurance that this machinery will be where it is needed, when it is needed. If this is not possible, and if importing it from other locations is not feasible, the process of the work sometimes can be revised.

Parallel projects

Another common drain on the availability of planned resources is the unexpected emergence of other projects to be executed at the same time with the same resources as the project that is being planned. For example, another utility may schedule a major outage, or a neighboring refinery may plan a capital expansion project for the same time period that this power plant project (or major outage) is scheduled. Since all too often there is little or no communication between the various industries in the locale, coordination of resource use seldom occurs, unless a backup plan is in place.

In this case, the effect on the availability of the supervisory staff is different from the previous case of project delays. Previously, the issue was how to keep key staff members available until the project actually started, with the knowledge that the remaining staff could be sourced from somewhere when the time came. In the case of two or more projects executing at the same time, the issue of availability of key staff does not exist; rather the availability of the remaining staff becomes the issue.

Usually, the project that starts first is the one that gets these remaining staff members, who are selected for their specific skills instead of their intimate involvement with the project. If the personnel with these requisite skills have already signed up for one of the parallel projects when this job starts, it does not mean others will not be available, it is just that they may have to be brought in from other locations, paid a higher wage to entice them to leave home, and also paid a per diem for housing, meals and transportation—costs that may have to be planned on during the preplanning stage of the work, that is, a contingency.

However, as before, the critical issue is the source of the craft labor. There is a fixed pool of labor in any given area. If the other projects in the area have the local labor committed to their projects, there may be a serious issue in executing the job at hand. If the job is labor intensive, importing labor from outside of the local area can become prohibitively expensive. Sometimes, a contingency plan may be needed to postpone the start of the job until the local labor is once again available. In the event of outage work, it may be more cost-effective to cancel the outage and plan to increase the scope of the next outage to include the scope from the canceled one.

Also, as with the issue of project delays, heavy equipment availability can be a problem when multiple projects are scheduled to execute in the same time frame. Similar to the supervision issue, the project that starts first usually gets the equipment first. In this case, there are several options. First is rearranging the sequence of activities to accommodate the equipment's availability schedule. Second, as in the case of project delays, the work process may have to be redesigned to exclude the need for the heavy cranes, excavators, and the like. But, there may be a third option. Sometimes, arrangements can be made with the management of the other projects to share the equipment. In other words, there may be a week or two when the refinery expansion will not need the large crane. If the project schedule can be rearranged to use that equipment during this time, the refinery management may be happy to sublet the equipment temporarily as a way to reduce the equipment rental costs for their project.

All of these issues need to be considered during the preplanning stage. Although it may seem premature to spend much time on these issues, they should be on someone's radar screen and periodically evaluated.

As a final suggestion, communication within the industry can minimize many of these issues. For unionized contracting work, maintaining a close tie to the union management, by both the labor-using contractors and the owners and plant staff, can lead to contingency planning that will help mitigate the problems of delays or parallel projects. Similarly, keeping in close contact with other contractors in the area and sharing information on projects planned and contracted is good intelligence that can be used for contingency planning. And last, never underestimate the information available from the equipment suppliers. They make many regular visits to the plant owners and contractors to learn exactly what work will be done today, tomorrow, and after tomorrow. They know that regardless of who will be performing the work, they will be supplying the equipment, and to stay in business, they must be prepared. Knowledge is the key, and one gets it by communicating.

Contingency dollars

However, no matter how much planning, questioning, and intelligence gathering is done, once the job gets the go-ahead, there will always be surprises. Therefore, it is very common for estimators, schedulers, planners, and management to add contingency money to a bid. The most common method is to add an arbitrary percentage to the bottom line in the belief that the money will be needed somewhere, sometime during the project's execution. Seldom is the contingency a calculated value based on specific risk analysis. As discussed in the paper "Contingency Misuse and Other Risk Mitigation Pitfalls,"

> Contingency is established to mitigate the adverse impacts of the unforeseen or under-predicted events. As such, contingency should be utilized and managed exclusively within the framework for which it is established. While a project budget document might contain several different "Fund" accounts as opposed to "Line" allocations, contingency is very different in that it is a reserve and "hedge" against risk.
>
> The manner in which Contingency Funding is developed dictates the guidelines of how it should be effectively managed. As contingency is risk-based, it should be sufficient to manage the realization of risks. The manner in which risk affects a project is a combination of constants and key variables. These will change relative to each other and to the Project itself at different points throughout the project.[5]

A monetary value should be specifically calculated to cover risk based on the possibility of it being needed as opposed to using a standard percentage. For example, if there is the possibility of needing to work through a holiday period, there should be a contingency amount calculated for the potential additional overtime instead of using a percentage adder. Or if the labor productivity is questionable due to possible disruptions from other contractors, a specific amount of additional man-hours should be calculated and included as a contingency. If a contingency to cover potential schedule liquidated damages is prudent, it should be calculated using a fixed number of people for a specified time frame. Continuing on from the previously quoted paper, "Contingency Misuse and Other Risk Mitigation Pitfalls,"

> Just adding an ad hoc ten to fifteen percent of the total budget for contingency poses two problems. The first problem is whether the line item contingency is adequate for the risks that are associated with the project. The second problem lies in determining when the contingency should be used. Holding contingency funds throughout the project and then looking for ways to spend the funds at the end of the project is not the most efficient use of project funds. This is especially the case where funds are limited and unused contingency funds could be used to fund other projects.[6]

Summary

A well-planned project starts with a preplanning process. This includes steps such as questioning the logic of participating in the project to begin with. It includes looking internally to see if the organization is prepared to move forward. Does it even have the resources to do so, and are the rewards worth the risks? The first step in this process is to assemble and fund a small preplanning team to answer those questions.

Next, it is important to do a lessons learned exercise. This should be done with participants who are expected to be assigned to the project as well as some who have experience with similar projects but have no stake in the one at hand. This group should think out of the box, so to speak. Utilizing the services of a facilitator can often be beneficial here. This group needs to look at all of the physical aspects of the project, such as weather, access, resources, and the commercial conditions. The group also needs to address social issues such as the project's impact on the community and how to involve those from within.

Even before a project is given the go-ahead to proceed, a delivery structure has to be considered to meet the organization's needs and expectations. Depending on the needs of the contracting organization, a variety of delivery structures are available. Some are based on guaranteeing price. Others focus on delivery time. Then, others focus on cost by placing the responsibility for risk management on the project participant who can most effectively mitigate it.

The volatile nature of the power generation business often makes it difficult for project owners to plan for the most effective project delivery structure very far in advance of the start of the project's execution. Just look at the number of projects that have been planned and then cancelled or switched to a different fuel. Therefore, the preplanning team sometimes must take a leap of faith and suggest a project delivery structure even before they have all of the facts that will impact the project, if it goes ahead.

From an owner's perspective, funding also takes center stage at the outset of a project. Where does it come from? How much is required? How is it to be guaranteed (to the lender)? An important segment of the funding involves the time of construction, because the longer this work takes, the more the interest costs increase. Sometimes it may be more cost-effective to double-shift or work overtime to keep the interest costs at bay.

Finally, no matter how thorough the planning, no matter how carefully and thoughtfully the plan was developed, certain things *will not happen as planned*. There may be labor strikes or bad weather. There may be bankruptcies. There most likely will be scope creep. There may be many other unforeseen changes, all of which must be accommodated in some fashion. Therefore, contingency planning must form a part of the overall planning process. The two most disruptive issues are project delays and unforeseen parallel projects. Therefore, some type of contingency funding, and management of those funds, must be developed to address these and any other potential disruptions.

Once this preplanning phase has been traversed, the next steps will be to develop the actual costs or numbers that the contractors will require to perform the work. For the owner, this means starting the bidding process, the subject of the next chapter. For the contractor, it means looking at all of the details and risks (the subject of the next few chapters) and learning more about planning (chapter 4). Read on for ways to answer many of the questions raised in the previous pages.

References

1 Lowder, Hoyt, et al. "Rooting Out the Problem." *FMI Quarterly*, issue 1, 2007, pp. 34–43.

2 Schimmoller, B. "," *Power Engineering*, vol. 109. October 1, 2005.

3 Bridgers, Mark, Clark Ellis, and Dan Tracey. "Buying Construction Services: Who Should Be Your No. 1 Pick," *Construction Purchasing*, Quarter 2, 2008, p. 22.

4 Bridgers et al.. "Buying Construction Services."

5 Noor, Iqbal, and Robert Tichacek. "Contingency Misuse and Other Risk Mitigation Pitfalls," paper presented at AACEi Annual Meeting, June 13–16, 2004, Washington, DC.

6 Noor and Tichacek. "Contingency Misuse."

Bidding 2

The bidding process is the time to head off many of the problems that could be encountered during the project's execution. A properly managed bidding process, followed by diligent project management, will go a long way toward ensuring a successful project. The trick to starting out right is to design and then manage the bidding process just as if it were a project in and of itself, because it is.

Before starting the bidding process, it is very important to ensure that adequate resources are available to prepare the bid in a timely and effective manner. It is important to view the bid with respect to the overall demands on the business at hand today as well as when the project will eventually execute.

Once the decision has been made to move forward with the bidding process, a thorough review must be performed of the specifications, drawings, and all other information available. If it is a complex project, the specifications should be separated into a logical grouping of disciplines and reviewed by experts in each field. One typical grouping could look like the following:

- Quality assurance and nondestructive examination (NDE)
- Commercial/legal
- Construction engineering
- Welding engineering
- Labor relations
- Safety
- Accounting (cash flow, taxes, credit worthiness, etc.)

During the review of the various sections of the specifications, it is advisable that the reviewers look at the project requirements from two perspectives: the owner's risks and the bidder's risks. The owner or author of the specification is usually looking for a proposal that is all-inclusive and with a minimum of exceptions, transferring to the bidder as much of the risk as possible. On the other hand, the bidder will be trying to minimize risk by taking prudent exceptions where risks are high but also placing a dollar or time value on those risks that can be mitigated accordingly. The ultimate objective should be to submit an offer that is a win–win solution for both the owner and the bidder.

However, before moving forward with the bidding process, a determination of interest must be made. The initiator of the request for proposal (RFP) should poll the market for an indication of interest. Then, a prebid meeting should be planned with qualified bidders. At this meeting, there should be free and open discussions to ensure that all bidders, and the initiator of the RFP, have the same understanding of what is expected. The scope should be discussed—and clarified. The same applies to the schedule.

Pricing should be discussed in very clear terms. Since the first indicator of contractor acceptability to the owner is price, pricing should be very transparent. The old practice of submitting a low-ball price with the intent to make it up with extra work should be circumvented. It serves no purpose in this day of tight resources and bottom-line focus. In fact, it can lead to lengthy, costly lawsuits. During these discussions, payment requirements should also be brought up and openly discussed, especially in terms of cash flow. If the project is going to have special tax treatment, this should be identified so everyone is on the same footing when preparing the offers.

Finally, penalties and bonuses should be discussed. Unfortunately, owners and contractors often have vastly different ideas of what motivates a contractor—a hefty penalty or a hefty bonus. As will be shown later in this chapter, there are ways to structure the RFP, and the resultant offer, that can incentivize the contractor and still save the owner money.

Managing It Like a Project

Do you want a successful outcome? Do you want your project to be a win–win for all participants? If the answer is yes, then the bidding process must be treated just like a project in and of itself. It cannot be overemphasized that most of the project's problems start right here, with the bidding process. Let's just take a simple thing like "assuming" that the requirements of this job are very similar to a previous one. This assumption can be devastating for the owner as well as for the contractor. If the owner essentially copies a specification from a previous job, yet it turns out that the terrain, the climatic conditions, and/or the skill sets for the new job are in fact not similar, the contractor will have cause to ask for (and most likely receive) extra time and money as the project progresses. If the contractor makes the assumption that the specification is essentially the same as a previous project, when in fact it is not, the contractor could be ineligible for additional time or money to make corrections. For example, the contractor who did not realize that the insulating requirements of the project were to apply two 2-inch layers of blanket insulation instead of one 4-inch layer, because he or she did not read the specifications in detail, could be required to reinsulate at the contractor's own expense.

So how can these misunderstandings be prevented? First, the owner must make a commitment to set up a specific process for requesting the RFP, and the contractor must make a commitment to set up a specific process for preparing the offer. Owners must know exactly what they want. This does not mean that things cannot be changed, but if they are, they should be changed procedurally. A specific set of written steps should outline exactly what items and issues need to be addressed during the preparation of the specifications. For example, a work breakdown structure (WBS) should be developed for the process, just as one would do for a project. Then, a schedule of when each step starts and stops should be developed. Next, assignments of responsibility for each WBS have to be made. And finally, the progress of the bidding process should be monitored to ensure a timely and effective completion.

Let's walk through a scenario, using the example from chapter 1 of building a 1,000-MW gas-fired combined-cycle power plant in a sparsely populated area where the summers can be severe. Let's first address the subject from the owner's perspective and follow up by addressing it from the bidder's point of view.

Owner's perspective

Two of the owner's initial items of importance normally are cost and schedule. How much will it cost to install the equipment, and how long will it take? The answers are often interrelated. For example, if overtime is included in the cost estimate, then it may be possible to shorten the schedule. On the other hand, if money is tight, then the schedule may have to be extended to avoid the overtime premiums. This often boils down to a matter of priorities. The 1,000-MW plant in our example is generally a long-term project. The cost of money during construction and the lack of revenue generation to repay it will form part of the original cost basis (loan).

However, as this book was being written, there was a significant shortage of skilled craftsmen and experienced supervision. Since this project is in a remote location, craft labor will need to be imported, housed, fed, and most likely trained. Supervision will have to be imported as well. And both will demand more than the standard 40-hour workweek—first for the extra money to entice them to leave home for the job, and second to fend off the boredom that could arise with all of the free time that comes with a 40-hour workweek. These issues must be taken into consideration as the owner prepares the specs and sets up the bidding process. Also, the owner will have to consider which contractors have experience working in remote locations. The owner will have to determine who will be responsible for material delivery and storage into and at this remote area. The time and the method for determining these factors will need to be included as part of the bidding process.

Then, to maintain control of the bidding process, the owner should list the steps (the WBSs) the process will require:

- Internal specification development
- Internal budget and schedule preparation
- Identification of potential bidders
- Issuance of RFP
- Prebid meeting
- Deadline for submittal of questions
- Deadline for response to questions
- Bid due date
- Bid evaluation completion date
- Bidder short-list evaluation
- Contract negotiations
- Contract award

Obviously, this list is not exhaustive. However, the owner should have a list similar to this, imposed on a timeline and updated regularly. This will help keep the bidding process moving forward.

To develop this timeline, or schedule, one normally starts with the need (or announced date) for the plant to begin generating power and works backward to the date the bidding process can start. In developing the schedule of the bidding process for the construction of the 1,000-MW power plant, one would start with the expected online date, move back in time for the duration of the estimated start-up period and expected construction span. Then, the mobilization period would be plotted as well as the period of time the contractors will need from notice to proceed (NTP) to mobilization. This would be the point in time, which could easily be 24-plus months for the 1,000-MW plant, that would coincide with the end of the bidding process. With this date then determined, the remaining time from the present until that date can be laid out. All of the activities listed above can be scheduled, including budget estimate, schedule development, specification preparation, RFP issuance and evaluation, and final award. As each step progresses, it and the total bidding process can then be managed toward their expected completion dates (see fig. 2–1).

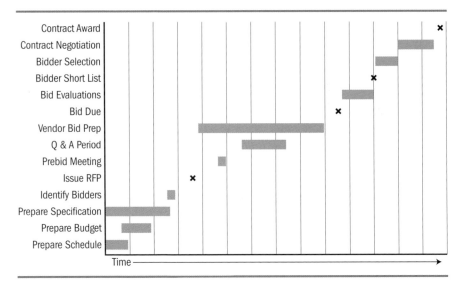

Fig. 2-1. Owner's bid process

To assist the owner in asking some of the questions that need to be asked, a comprehensive list, similar to the one in Appendix A, "Owner's Construction Estimate Checklist," could be developed. This list is not all-inclusive, and it applies to more than just the development of the bidding process. It is also useful for the estimating or budgeting phase, which will be described in more detail in the following chapter.

Bidder's point of view

While the owner has the responsibility of getting the ball rolling, the bidders have a major role as well. They have to evaluate and understand what is being asked of them, and they do not have the luxury of setting the bidding schedule; as discussed above, this is normally dictated by the owner. The bidders have to put together their own bidding process, and it is usually more complex than the estimating process used by the owner. Usually, they are not privy to the owner's internal policies and decisions regarding the specific project, which often precludes them from making a bid/no-bid decision until after they have spent time and money reviewing the bid specifications and visiting the site. If for no other reason than this, there are many in this industry who recommend that consideration be given to some kind of partnering arrangement, as discussed earlier in chapter 1.

Therefore, the contractors planning to bid for work such as the construction of a 1,000-MW power plant, or portions thereof, must consider a host of issues, and they must all be reviewed within the bid time frame provided by the owner. The following, which are included in Appendix B, "Contractor's Construction Estimate Checklist" are some of these issues:

- Quality assurance review
- Commercial and legal review
- Rigging, welding, and other specialty processes review
- Labor relations review
- Safety issues review
- Job site administration review
- Cash flow review

Each of these issues often requires input from different departments within a contractor's organization. To coordinate and manage them can require a sizable effort. Contractors that bid for work on a project such as a 1,000-MW power plant are often bidding on other work at the same time, as well as executing a variety of ongoing projects scattered around the country and globe. In other words, these organizations are usually very busy. Scheduling reviews of sections of new specifications is not always easy. The reviewers frequently have their priorities already established, and the current project may not be part of them.

However, if these reviews are not made by the parties with the best expertise, the bid may be jeopardized. The bidder may unintentionally ignore certain owner requirements, which could cost the company if it is awarded the work. Or, the bidder could add contingency money to cover an "unknown" and then not be the selected contractor for the work due to the higher price with the contingency. Which is to say the bidder would have put forth a lot of effort and spent a lot of money to bid for work that it was doomed to lose. Neither situation needs to happen. A structured, preplanned bidding process will avoid this.

First, potential bidders will need to decide if they want to bid. To do this, they will need to ascertain if the work fits with their companies' needs and goals. Assuming that is the case, the next step is for the bidders to determine if they have the necessary resources available to prepare an effective bid, in the time frame allotted by the owner. These two issues will be at the heart of the bid/no-bid decision. If this is a go, then a detailed bidding plan must be developed, personnel must be located, and a schedule of bidding activities must be agreed upon. This bidding schedule will look something like the one in figure 2–2.

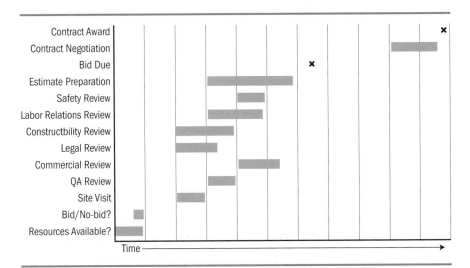

Fig. 2-2. Bidder's bid process

As with the owner's bidding process, the bidder must also monitor the bid process to ensure a timely and effective completion. For a project like the 1,000-MW unit, weekly reviews should be conducted, and authorizing management should be kept informed of the progress of the bidding effort. Each discipline involved in the bidding process should be encouraged to complete their review and provide support to the bidding team to avoid a crunch effort during the last days of the bid preparation. Many times, questions arise at the last minute that can create an atmosphere of near panic, such as code applicability, labor and equipment availability, risk, contingency, and so forth. Not keeping authoritative management informed can lead to snap decisions at the last minute before the bid is due to be delivered. These snap decisions may add unnecessary costs to the bid. They may result in overlooking certain risks or in a no-bid decision after weeks of effort have been spent preparing the bid.

For example, the 1,000-MW unit being bid for installation in a hot, barren area of the country will most likely have labor availability issues. If not enough time is allowed for the labor relations designee to talk with the unions, or the purchasing group does not have adequate time to talk with subcontractors, authorizing management may add a sum of money to cover the eventuality of needing to import and house more labor than will be required in reality. This could easily escalate the price of the bid and take the bidder out of consideration by the owner. Or, if not enough time is given for constructability reviews, certain difficulties could be missed. For example, the constructability reviewer might not realize that there will be limited access for the transport and setting of the HRSG (heat recovery steam generator). Without adequate time to make a site visit, the reviewer may not realize there is an existing structure in the way or an old riverbed that needs to be bridged. This could result in costs that are not included in the estimate.

With a proactively managed bid process, however, these types of issues should be minimal. There will be time for verbal agreements to be documented. There will be time for a site visit. There will be frequent question-and-answer meetings where concerns can be addressed in a timely fashion. The final days before the bid is due will not be spent rushing around, looking for answers and approvals. The proverbial midnight oil will not have to burn as often if the process is properly monitored and diligently managed.

The Specifications

The specifications for the work are always at the heart of the bidding process. They spell out what is expected, and what it is that the client wants or is trying to accomplish. However, they are often interpreted differently by the different parties to the project. For this reason, it is important that they are clearly written, without the use of ambiguous words, and thoroughly vetted among the peers on the project and possibly third-party experts. For the project to be successful, the specifications have to be clear and concise. They must address all aspects of the project, either directly or through references.

Being the author (how to write a spec)

Writing a specification runs the gamut from very easy to extremely complex. If the project is a repeat of an earlier project, many sections simply can be copied from the specs or contract of that earlier project. If the project is a large, new undertaking such as the 1,000-MW unit referenced earlier, the task becomes more daunting. In this case, there may be a person or persons dedicated full time to preparing the construction specifications. Other groups may need to be brought on board to prepare some of the sections. Some parts may be best left out, and instead be just referenced, for example, code requirements.

A good specification gets started with an outline. A good specification outline for a construction project such as the 1,000-MW installation project should include most of the following:

- Project description and scope
- Schedule and constraints
- Price and price changes
- Payment requirements
- Taxes
- Penalties and bonuses
- Terms and conditions (the commercial rules)

The authors of the specification first collect all of the information necessary to prepare a quality document by interfacing with many people. They will talk with the designers of the equipment to be installed and the purchasers of the equipment. They will spend time with the operations side of the business and be in touch with the legal department, as well as many other groups. But a group that is seldom brought into the spec development process is the group of people who will be preparing bids in response to the specifications.

Although the persons preparing the specifications may have prepared many specifications before, they have usually done it the same way each time. They seldom spend much time listening to the bidders and the difficulties they may encounter in trying to respond to the requirements of the specs. This is a mistake. The authors of the specifications should spend time with the prospective bidders before finalizing the specs. They should solicit ideas from these prospective bidders, ideas that will make everyone's tasks easier. For example, a typical owner's specification may call for the bidder to provide/prepare/install whatever else needs to be provided/prepared/installed to complete the work "in accordance with industry standards." That is an open-ended requirement that is subject to as many interpretations as there are interpreters. This usually arises when the author of the specifications, say, the owner, has always used the same contractors and the same contract documents and has developed an informal understanding with the contractors of what this phrase means. However, there will be times when either there is a new contractor, or there are new managers with the same contractor, who are not privy to the past relationships. They may not make the same assumptions that the former parties made.

Asking for input from the potential bidders will go a long way toward avoiding misunderstandings. These bidders could suggest alternate wording that they have previously used, or they could offer clarifications that would prevent future conflicts. Although some say that this could be done during the prebid meeting, by the time the prebid meeting is held there is precious little time before the bid due date to make significant changes to the specifications. And if nothing is done before the bid due date, the bids will most likely come in with varying clarifications, or worse yet, no clarifications. Varying clarifications will create difficulties when the owner attempts to evaluate the bids. If there are no clarifications, the contract may be open for varying interpretations during the execution of the job, which often leads to conflicts.

The author of the specification is therefore the first gatekeeper of future problems. The less ambiguity in the specification, the less the likelihood of future misunderstandings. As first addressed in chapter 1 doing a bit of group-think or conducting brainstorming sessions will go a long way toward ensuring a well-prepared document. Including the end users of the installed facility, as well as the potential installers in these sessions, makes this process even better.

Being the bidder (how to read a spec)

OK. The specifications have been written and they have been issued. Now it is the bidders' turn to deal with the specs. They have to familiarize themselves with the specs and decide what is important and what is not. The specs for a large construction project like the 1,000-MW project is more than one person can reasonably expect to understand and bid on in the time frame normally allowed by the owner. Yet it is crucial that the bidder thoroughly understands the complete set of bid documents. Envision a comprehensive set of specs produced by the owner or architectural engineering (AE) firm. These specs will consist of many, many documents relating to each major area of the plant. There will be a set or group of specs for the site preparation works; another set for the site civil works, sometimes including much of the above ground structural steel works; large specs for the HRSG island, the turbine/generator island, the cooling system, the chimneys, the plant electricals, the instrumentation and controls systems, and the control room itself; and so on and on. Ultimately, there either will be reams and reams of paper and drawings for all of these areas or digitized specs that equate to the reams and reams.

Any one bidder usually will restrict involvement to only one, two, or three areas of the plant. Possibly, the bidder will only be interested in the boiler island, and if it is a coal-fired plant, also the coal conveying system and the ash removal and environmental equipment. Other bidders will then be interested in the turbine island, the cooling systems, the site preparation works, electrical, and so forth. But whichever system bidders are interested in pursuing, they must understand that they are not each working alone. They need to view the work scope as an integral part of the total plant construction process. And they must also visualize the end product being integrated with all of the other power plant systems.

Let's examine this a bit further. As the bidders are reviewing the specifications for a particular work scope, they will have to envision work crews from other contractors moving around and even sometimes through the areas where they will be working. The bidders will need to consider not only the production effects of these types of disruptions, but also any potential safety issues that could arise. They will need to plan on coordinating efforts with some of these other contractors. For example, for a coal-fired boiler, if the boiler steel is being erected by Contractor A and the boiler itself is being erected by Contractor B, there are periods of time when it will behoove both contractors to coordinate their work. Contractor B may ask Contractor A to leave out certain steel so some of the boiler components can be installed before completing the steel makes access difficult. In the same way, Contractor A may ask Contractor B to accept the responsibility for installing grating, handrails, and some other structural components that were left out to prevent damage by Contractor B's workers. The same will apply to underground utilities and electrical cabling; this work will need to be coordinated with all of the other contractors in the area.

Bidders must also visualize the overall end product—the total plant. Since the owner's objective is not just to install equipment, but rather to have an operating plant that will generate electricity, all of the pieces of equipment that will be installed must also work together to achieve this. Therefore, the bidders must consider equipment and component compatibilities. They must consider working together with the other contractors to integrate the start-up procedures. If they are bidding for installing the balance of plant (BOP) equipment, they need to understand that the boiler feedwater pump, including its motor, motor control center, cabling, and piping, must be in working order by the time the boiler contractor is ready for performing the boiler hydrostatic test.

In the same way, the boiler erection contractor has to coordinate with the turbine/generator erector. There will be a specific time in the construction schedule when the turbine erector will be ready to roll the machine for the first time. The boiler erector has to be ready to provide steam at this time, steam previously proven to be clean to a certain standard. If the bidder does not have this in view when reviewing the specifications, and when preparing the bid, this may lead to conflicts later on.

When reviewing the specifications for the work in their area of interest, bidders must mentally walk through the entire erection process. As they do this, they should be preparing a list of questions to be asked and issues to be clarified at the prebid meeting. Since it seems that there is never enough time to prepare a bid, it is essential that the specifications be thoroughly reviewed very early in the bidding process, and the bid schedule that is developed for the bidding effort should reflect this early review.

Additionally, bidders must be considering the commercial risks they are being asked to accept. These can run the gamut from warranties or guarantees to penalties or bonuses and from cash flow to tax issues. Once a contract is signed for the work, it is usually difficult to make changes that shift risk or costs from the successful bidder back to the owner.

As shown in figure 2–2, bidders should separate the specifications into sections similar to those shown in this schedule. Always keeping the bid due date in mind, bidders should parcel out these separate sections of the specs for review by the persons responsible for these areas within their organizations. If there is no individual specifically responsible for an area, consideration should be given to utilizing a third party for help. In today's world where litigation often trumps cooperation, not being thorough, and thereby not clearly understanding the risks that one is being asked to accept, can lead to disaster. The following points out the consequences of this; it is from a real-life situation:

A small contractor, without an in-house staff to review all of the commercial requirements, chose to gloss over the detailed clauses of the specifications, some of which pertained to payment of costs incurred in the event of delays. Specifically, the specifications stated that if the contractor was delayed in the completion of the work due to causes beyond the control of the contractor, the time for completion would be extended. The specifications then went on to say that such an extension would be the contractor's *sole* remedy. That is to say, extra time would be granted, but not the costs incurred thereby!

Elsewhere in the specifications, there was another clause stating that claims for any change in the contract due to hindrances *by others* would be made for time only. In other words, no recovery of additional costs would be allowed, even if the fault was due to actions, or inactions of others—such as late releases of areas to work in!

Since the clauses mentioned above were buried within a couple of paragraphs of a very voluminous specification, they could have easily been missed as a concern by a bidder not staffed to note such wording. In the case of this bidder, who was awarded the contract to perform the work, this was missed. The job ultimately ran over schedule by many months, in large part due to late releases of areas to work in. The contractor was not paid for the extra costs incurred. Bankruptcy followed, and lengthy, costly lawsuits were filed. All could have been avoided if the bidder had in-house staff, or even an outside third party, make an astute review of the commercial terms of the specifications. The bidder could have then either negotiated more reasonable terms or added contingency money to the bid. In this case, the bidder did none of these things and thus suffered the consequences.

Similar issues can arise from an inadequate review of the technical section of the specifications. Another specific example follows:

The specifications spelled out that the ASME (American Society of Mechanical Engineers) code edition and addenda used for the design of the boiler also were to be used for the fabrication and installation of the boiler. In the past, ASME codes always required that a boiler be installed in accordance with the edition of the code to which it was designed and manufactured. However, starting with the 2013 edition, a boiler can now be installed in accordance with either the edition of the code used for the design of the boiler *or* the edition in effect at the time of installation.

In this case, the installation work that was now under consideration was being bid on three years after the unit was designed, and the codes had been revised during that time. Specifically, the code edition in effect when the unit was designed required certain welds to be postweld heat-treated. The 2013 edition does not.

The person reviewing the specifications was keenly aware of the revised code requirements and therefore did not include money for stress relieving these particular welds. The people in the field also knew the code no longer required these welds to be stressed relieved. So they were not.

However, the general contractor (GC), upon review of the contractor turnover packages, did remember that the specifications required the unit to be installed to the older ASME code edition. The contractor was required to rescaffold, remove insulation, and stress relieve and reinsulate all of the affected welds. Not only was the contractor not able to recover the costs for the stress relieving that was not priced into the work but also not able to recover the costs of having to return to the area, rescaffold, remove and replace the insulation, and so forth. A high price to pay for not understanding the work and the specifications for it.

The opposite could also happen. Since the requirement to construct a boiler in accordance with the code under which it was designed was only eliminated starting with the 2013 edition of the ASME code, bidders who are not familiar with this new aspect of the code could potentially add money into their estimates for work that is no longer required, for example, the above postweld heat treatment. In summary, whether one is the preparer of the specifications or the bidder, clarity is paramount.

Scope

Scope is what the job is all about. It is the product or service offered by the seller and accepted in writing by the buyer, so it must be as clear as possible. The person preparing the specification and the person reviewing the specification and preparing the bid must both understand the scope in the exact same way. On the one hand, if the preparer of the specifications for a major condenser retubing job had intended for the successful contractor to also repair the water boxes, if they needed repair, and if this was not made this clear in the specifications, the bidder may not have included the eventuality of this part of the work in the bid price. This, obviously, would lead to a disagreement on the site. On the other hand, if the bidder was a contractor experienced in this type of work, the bidder should have asked if the repair of water boxes would be expected, and then prepare the bid accordingly. The key here is to be extra clear, even to the point of being redundant.

Another area of scope delineation is where one contractor's scope ends and another's begins. Take the case of installing the feed water line to the boiler economizer. Let's suppose Contractor A is responsible for the installation of the feed water line to the boiler while Contractor B is responsible for the boiler itself. The question that frequently comes up is: Who is responsible for the final weld between the feed water check valve and the boiler economizer inlet header? Drawings usually show a line at this point with two arrows pointing in opposite directions and a note that says: Piping work stops here (Contractor A) and boiler work stops here (Contractor B) (fig. 2–3). However, it does not always indicate who has the responsibility for the junction weld itself.

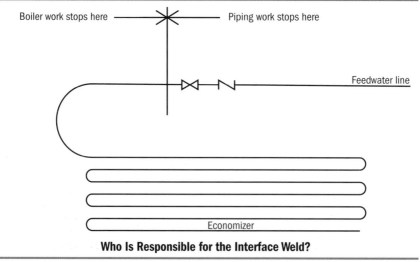

Boiler work stops here ——————— Piping work stops here

Feedwater line

Economizer

Who Is Responsible for the Interface Weld?

Fig. 2–3. Who is responsible for the interface weld?

Scope issues are related not only to the materials to be removed or installed but also to support issues such as trash removal, road maintenance, the supply of construction power, water, and so on (see appendix C). If not clearly defined, or agreed upon shortly after moving on-site, costs can be incurred that may be difficult to recover later. For example, the scope section of the specifications may be perfectly clear that the owner is responsible for the supply of construction power. However, if the author of the specifications does not spell out to what point the power will be supplied, the bidder may just automatically assume it will be provided to the distribution board, at 240 V and 120 V. This would suggest that the owner is responsible for the step-down transformers and connection points. However, if the owner had only intended to provide a connection upstream of the distribution board, the bidder would either be short of money for this extra equipment, or the bidder would be going for an extra to the contract at the very beginning of the job. Had the specifications been clearer, or had there been included in the specifications a document similar to appendix C, this would not have happened.

Another scope item that is often not clear in the specifications is touch-up painting. Let's assume that Contractor A is bidding to install the boiler feed water line. As part of this scope, Contractor A would also be responsible for the hanger support system. But to install this system, this contractor first would have to erect some steel structures and platforms, which would require welding and bolting to the existing boiler support structure. In the course of this work, however, Contractor A's workers will damage the paint of the existing steel. The question arises, who will repaint, or touch up, these areas? Who should provide the paint (to match the existing paint) and who should provide the labor? The specifications must be clear on points like this.

Schedule and Constraints

In addition to being clear on scope, the specifications also must be clear on the schedule and any potential constraints therein. Generally, the schedule included with the specifications would spell out only the critical project dates such as initial access to the site, some intermediate dates of material deliveries, and other contractor interfaces, and then the dates for testing to begin and when the work is to be complete, ready for turnover to the client. Sometimes this will be in the form of a list of dates only; at other times it may be in the form of an actual schedule, ready for insertion of the details by the bidder. But whatever form the schedule takes, it will almost certainly have financial ramifications if the activities within it do not meet the stipulated dates. Therefore, bidders must be very clear about accepting these dates. Bidders must also be clear about any dates or durations they provide.

As mentioned earlier, in most cases the successful bidder will be only one of several parties engaged on the project. The commitments made during the bidding stage will carry over to the other bidders. For example, if the bidder for the major civil works accepts a clause to complete the turbine pedestal by a certain date, then the bidder must also be prepared to accept serious penalties in the event it is not completed on time. And depending on how tight the project schedule is, these penalties could be extremely high; a few hundred thousand dollars a day are not uncommon.

For the owner or GC, it is imperative that all parties are in agreement on the form of the schedule, usually some type of a critical path method.

The software used to produce the schedule is not as important as the format, but if agreement can be reached that all parties use the same process, then updating and exchanging data will be greatly simplified. As more sophistication is introduced into the construction industry, more and more contractors are being asked to plan on using very specific software and formats that can be integrated with the other contractors' schedules, and then rolled into one overall project schedule.

Far too often, there end up being several schedules on-site. The owner may have one schedule, which is updated based on input from the contractors and an analysis by the owner's staff. The general contractor usually has a schedule that sometimes is not linked to the owner's, and the various subtier contractors sometimes also have their own independent schedules that are not linked to each other or to the general contractor's. This is a recipe for disaster.

When different parties, contractually bound to work toward the same end goal, do not use one consistent master schedule, deviations from the critical path will develop. Differences will arise in the reporting of stages of completion of the various activities from contractor to contractor, and paths that should be integrated to ensure getting to the completed state will no longer be integrated. Frequently, this happens because each party does not want to divulge to the other everything they know about the job. Sometimes this is due to potential or ongoing claims and litigation. At other times it has to do with payment that may be adversely affected. The ultimate risk is that the project momentum loses efficiency because everyone is not pulling in the same direction.

In the best interests of the project, it is important for the specifications to ensure that the bidders understand that they will be required to work together using a single master plan. It is important that all parties understand the constraints that will be placed on each other if they do not meet their expected commitments. To illustrate this, let's look at the following example: Say the electrical contractor saw an opportunity to start installing cable trays into the boiler steel structure four weeks early. This contractor proceeds to install the trays, even pulling some cables, and then removes the scaffolding and cranes used for this work and reports the activity as complete—maybe even getting paid a substantial sum for activity completion. One week later, the boiler contractor moves in to start installing the reheat elements and cannot get the elements into the boiler cavity because the cable trays are blocking the access (fig. 2–4). Obviously, the cable has to be pulled back out and the trays removed. Who reimburses the electrical contractor for this removal and reinstallation? Who reimburses the boiler contractor for the delay mitigation while waiting for access to install the reheat elements?

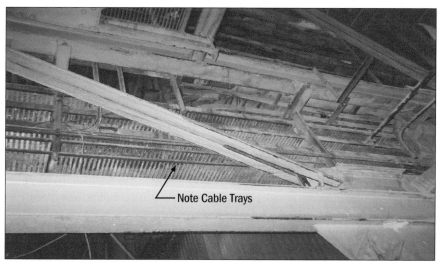

Fig. 2–4. Cable trays installed prematurely, blocking access of installation of reheat elements (Courtesy of Construction Business Associates)

Unless the *constraints* to the schedule are understood as well as the basic schedule itself, problems will occur. As just seen, extra work will be done by the electrical contractor with little chance of financial recovery. Delays that never needed to have occurred will be incurred by others. To minimize the chance of problems such as this, the specifications should be clear that regular (daily?) progress updates must occur, and the updates must be to a single, master project schedule to which all participants have full access and a clear understanding of its logic and rationale.

Let's illustrate how the lack of properly specifying the use of a single fully integrated master schedule could hurt an owner whose daily lost revenue could easily equal $600,000 or more. Assume a major plant overhaul outage with some work, like the boiler, being contracted, while other work, like the BOP, is being performed by plant labor. As the work progresses, the contractor may be working to a detailed, sophisticated schedule, but the plant personnel often are not, as they are moved on an almost daily basis, from task to task, sometimes not completing one task before being "temporarily" moved to another to accommodate other plant issues.

Subsequently, envision a delay in the start of the boiler hydrostatic test, because the feed water pump was still disassembled. The plant personnel, who had been working on this pump, have been reassigned elsewhere. Although the overall project logic would say that the hydro test should not be scheduled to start until after the feed water pump was ready for service, here is a case where the boiler work schedule and the BOP work program have not been coordinated. If the specifications did not require that a formal link be established between the BOP work and that which was contracted to third parties, this type of problem can happen.

Price, Price Changes, and Qualifiers

Price

The contract price is the amount of money to be paid by the buyer to the seller for the product or service being sold. Every specification should include a requirement that a price or pricing structure be provided that is *directly* related to the scope and schedule. The specifications should also require that the successful bidder acknowledge that price, or pricing structure, upon award of contract. Also, whether or not extra work is anticipated, a request for rates and/or unit prices for such work should be included, and possibly standby rates should as well.

Price changes

But as we have seen, issues do arise. They may be related to scope changes. They may be related to support services required. They may be related to scheduling issues, or they may be due to misunderstandings resulting from a specification that was interpreted differently by the different parties involved. Knowing that we do not live in a perfect world, a well-prepared specification and a thorough bidding process will address not only the basic scope of work but also these eventualities and how to deal with them, up-front.

But what happens if extra work exceeds the amount of work the project is structured to manage and administrate? A well-structured contract will limit the percentage of extra work that can be performed under the extra work rates. A typical example of this limitation is a clause stating that if the extra work exceeds 15% of the base contract value, then the contract price shall be subject to renegotiation. This is done to protect both parties. On the contractor's side, if the work significantly increases, the contractor may not have adequate supervision and site support facilities in place to handle this increase of scope and, without this limitation clause, may not have any recourse to be reimbursed for these extra costs. On the client's side, if the scope is reduced significantly, and this type of clause does not exist, then the client may be paying the contractor for establishment costs originally anticipated but now not needed.

Qualifiers

Then, there also may be a need for including qualifiers to address special circumstances and specific assumptions made when the bidder developed the price. For example, the bidder could include qualifiers, such as the following, that spell out what is expected to avoid potential problems from issues such as the delays incurred in the previous feed water pump example: "The Seller will meet the schedule guarantees *provided the Buyer grants unobstructed access to the site, with the equipment prepared, ready for use, on the date agreed in the contract schedule.*"

Another qualifier that helps clarify intent is the following: "The Seller will meet the schedule guarantees *provided the Buyer provides all materials to the site, prepared ready for installation, in accordance with the installation sequence.*"

To illustrate the importance of the second case, let's look at the problems an insulation contractor can run up against. Like most projects, assume that this one has an immovable date of completion. Assume that the ultimate contract does not have the second clause referenced above, but that it does have heavy penalties for not meeting the completion dates and no other "changes" clause that addresses relief of schedule. (As already noted above, this should not happen, but sometimes it does.) The main contractor now schedules a set of work release dates for the insulation contractor to get access to the work areas. Now suppose the material arrives ahead of schedule, but as the work area release dates arrive, the mechanical contractor has not completed work sufficiently for the insulation contractor to get access as originally scheduled. This obviously prevents the insulation contractor from executing the work per the agreed-upon plan.

Finally, the mechanical contractor does complete the work, the insulation contractor does get total access, but what was originally scheduled to take only seven months to complete now requires 11 months from start to finish. In fact, as can be seen in figure 2–5, 50% of the area was not even released until the seventh month, so the insulation contractor completes four months behind schedule. The main contractor, however, cites the insulating contractor for failing to meet original progress dates and imposes liquidated damages. The insulating contractor sues for extra time and money since it was required to stay mobilized longer than originally scheduled, through no fault of its own. They go to court to fight it out when this could have been avoided with just the simple clause above included in the contract.

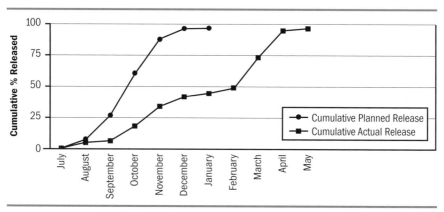

Fig. 2–5. Impact of late releases

But what happens when the issue is not directly related to schedule dates and instead is related to scope creep? Staying with the same insulation, mechanical, and main contractors, let's look at this problem. Suppose the insulation material was provided directly by the main contractor to the insulator, free of charge. When the insulation contractor prepared a price, this contractor was told that it would be responsible for insulation installation over a certain area and informed about the quantities of material that were to be installed; the insulating contractor prepared the price accordingly. However, time passed, and as the design progressed, the size of the equipment to be insulated grew; for example, the ductwork increased in size. No mention of this was made to the insulating contractor as it prepared to mobilize. It was only well after the job was underway that the insulating contractor realized that it had a significantly greater amount of material to install than it originally bid.

Unfortunately, the insulating contractor's contract, which reflected the specifications, did not address scope creep. The main contractor took the position that the insulator should have reaffirmed the quantities before starting work, since the contract did clearly state that it was the insulator's responsibility to complete the whole job without additional compensation. The insulator stated that its quotation clearly reiterated the quantities it expected to install. Short of arguing the points of law in this case, both parties felt strongly about their positions, and court was the only option.

Here again is a classic case of not thoroughly reviewing the specifications and not thinking through various problems that could arise, especially by the insulating contractor. Depending on the severity of the changes in the two cases above and depending on the honesty of the parties involved, the court may find for either party—something that would not have been necessary if more thought went into the review of the specifications and the development of the contract document. And changes go beyond just schedule and scope. They also can include weather conditions, strikes, acts of war, lost shipments of materials, and a host of other things that will be covered in later chapters. One cannot overemphasize the necessity of addressing the mechanisms to be employed in the event of changes, because *they will occur*.

Payment Requirements (Cash Flow)

The name of the game in the construction business is almost always "cash flow." Without an adequate cash flow, labor cannot be paid, suppliers shut off their credit lines, and the job comes to a halt. The first line of defense against this scenario is an understanding of the importance of cash flow at the bid stage. When the specifications are being prepared, thought needs to be given to cash flow. If the project is a large, new construction project like the 1,000-MW unit discussed earlier, most of the construction funding is usually in place by the time of the bid stage. In that case, the owner or AE can specify that the contractors will be paid in a fashion that keeps them cash positive, because the money will be available. However, if the project is a major rehabilitation project, with areas of unknown scope, cash flow problems may occur. Suppose that when the turbine is opened, severe damage to the blades of one of the stages is encountered. This could cause unexpected costs as well as schedule delays for the outage, and money might not be readily available for these extras. Then, an issue of funding responsibility may arise.

Cash flow—funding

The owner or GC needs to consider the available funds in line with the total potential scope of work. But so do the bidders. The bidders need to understand what costs they will incur and relate those to the payment formulas that the specifications allow. And they need to be prepared for handling extra work or cost overruns. During the bidding process, the bidders need to prepare a cash flow curve to ensure that (a) they stay cash positive or (b) they arrange for financing to carry them through the cash-negative periods. See figures 2–6 and 2–7.

Figure 2–6 depicts a contractor's potential cash flow with only progress payments (minus retention), as the job physically progresses. Note that from the beginning, Period 1, the contractor's cash out exceeds the cash in from the client, and it stays this way until the end of the job. Only after the job is complete, and the retention is paid (not shown on the graph), does the contractor's cash flow become positive. If this is how the job is bid and the contract is ultimately structured, this contractor will have to either self-finance the shortfall or borrow the money. Either way, it is a cost that must be considered during the bidding stage. Figure 2–7 shows what can be done to mitigate this circumstance. It shows the exact same $50 million project, but with a mobilization fee being received during the first period of the project. The contractor in this scenario is essentially cash neutral for the first half of the project, and although the cash flow does go negative for a few periods, it recovers shortly thereafter to a cash-positive position (less retention) before the job is completed. This is much more preferable, and each bidder for a similar project should strive to obtain this, at a minimum. In the case of these two scenarios, to maintain a cash-neutral position would require the client to provide approximately 3.5% of the funds up front. This is generally less costly for the client than asking the contractor to finance the work.

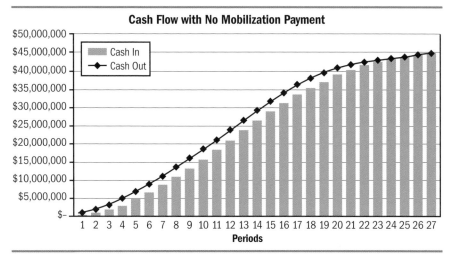

Fig. 2-6. Cash flow without mobilization fee

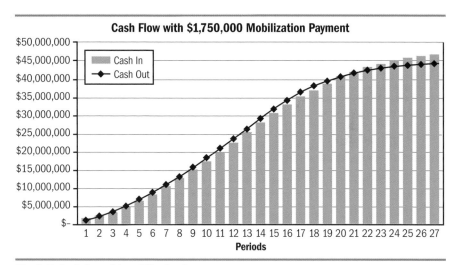

Fig. 2-7. Cash flow with mobilization fee

Cash flow—receiving

Similar to understanding contractual commitments such as scope, schedules, and how to handle changes, it is extremely important that the specifications and subsequent bids are clear on how progress and/or milestone payments are triggered. Once the work is underway, the site staff must understand what the contractual requirements are so they can get paid for the work performed. There can be any number of systems of payment releases triggered by just as many different events. A typical contract may have the owner providing an up-front down payment of, say, 10% of the anticipated contract value. Then, the next 80% may be tied to either percent progress or to milestones reached, or sometimes a combination of both. The last 10% will usually be held in abeyance until the unit meets certain operational criteria such as a successful 100-hour run. But it is normally the on-site personnel who will "pull the trigger" for the payment process to start or not start, based on how the contract is structured.

Normally, the contract should spell out exactly what events must happen for anyone to claim a payment. It also generally should spell out the process to follow to process the claim for payment. For example, the contract may call for a 10% mobilization fee upon successful completion of establishing the site offices, tool room, and changing facilities. It may further spell out that the next 20% will be paid upon completion of the inspection of the turbine internals. This is a total of 30%. Then, there may be a progress payment sequence for the next 50% of the project, starting with the repair of components and their subsequent replacement. Finally, another 10% may be triggered when the unit is handed back to the plant for synchronization, and the last 10% may be held until six months later, or after a certain number of hours of successful operation.

The important thing to realize is that only the on-site staff will know when the payment milestones have been reached. Therefore, the contract needs to be clear on how they are to initiate the process that generates an invoice. Since cash flow is so extremely important, the bidder should include money for someone who is intimately familiar with the payment sections of the contract to be at the site.

Sometimes there are also bank criteria such as letters of credit that have very specific requirements that must be met before funds are transferred. Maybe partial lien waivers are required waiving the contractor's rights of attachment for all work performed through the date of the invoice. This needs to be understood by the bidder, and the bid and cash flow should reflect any time delays due to this additional process. Again, it is all about cash flow, and the prudent bidder should leave no stone unturned to ensure that the payment process is designed for a positive cash flow.

Taxes

The subject of taxes is very complex. There are people who do nothing but work with tax issues to ensure that (1) no laws are broken; (2) all regulations such as payments and recordkeeping are followed in a timely manner to avoid penalties, which can be very large; and (3) full advantage is taken of all benefits allowed by the tax rules. Record keeping is very important. Sometimes, the taxing authorities ask for verification that their requirements have been met. They may perform audits to ensure compliance with their rules and regulations. They may require the business entity to prove that third-party audits have shown the process to be in compliance.

On construction projects, there are a host of tax issues that must be taken into account. The first and most important are the payroll taxes. Every jurisdiction, whether local, county, state, or federal (and even beyond our borders), has requirements to ensure that the workers and their employers pay the prescribed amount of taxes. These jurisdictions will also require the payer to provide verification of wages paid and taxes withheld from the employee, and this must be done by a certain date— usually by issuing a W-2 form if the work involves U.S. citizens or residents. Almost all contracts between the owner and the contractor, or between the contractor and the subcontractors, require that all statutory require- ments be met, especially in the area of taxation. This is done to protect the contracting party from having the taxing authority attach the site for future sale because someone did not pay their taxes. So once again, good record- keeping, in addition to withholding the correct amounts from employees and paying taxes to the taxing authorities is of paramount importance.

Other taxes that are unique to construction sites are for the purchase of gasoline and materials that will form a permanent part of the structure being built. Although not all jurisdictions follow the same rules, generally the taxes on gasoline are exempt if the gas is being used in vehicles that remain on-site. For a two- or three-year major power plant construction project, this can be significant. Materials that will form a permanent part of the final structure are also often exempt from local and state sales taxes. If the project is large, and if most assembly work is done in the field as opposed to in the manufacturing facilities, then there may be major dollars at stake.

The prudent bidder will research the statutory rules that govern these situations and then structure the bid and contract accordingly. For example, if large quantities of gasoline or diesel fuel are expected to be consumed solely on-site, the installation of a fueling station may be a very cost-effective move, and the bid should reflect this. Other taxes such as those on profits and imported goods and services also must be addressed at the bidding stage. For example, is the project exempt from local sales tax, due to special considerations for attracting it to the area? If so, the owner or GC should make this clear in the specifications, and the bidders should separate the taxes from the other as-bid numbers.

Penalties and Bonuses

A distasteful subject, penalties are a way of life in the contracting business. The majority of the contracting entities are still working from the days where compliance with contract requirements was thought to be best enforced through the imposition of penalties for noncompliance. Although this does have the intended effect, it is not necessarily the most cost-effective approach to contractual satisfaction. Usually, the contracting party's specifications will impose some type of financial deduction against the payments for the work if certain milestones or performances are not met. Sometimes these penalties are calculated based on perceived damages to the client but more often than not, they are just an assigned value that the client takes right to the bottom line, if imposed. The term used for this is liquidated damages (LDs), which essentially means that in lieu of calculating the damages incurred, this assigned value will cover whatever the cost may be, and both parties will accept this imposition.

Since these LDs are generally imposed on a calendar day basis, they could add up to very large sums if not properly managed. Therefore, they are usually capped at a maximum. This limit normally is first specified in the bid documents and then possibly negotiated before the final contract language is accepted, either as a specific dollar value or as a percentage of the contract value.

As pointed out earlier, a unit off-line can prevent an owner from realizing revenues of up to $600,000 per day. Certainly, the owner wants to recoup any of this unrealized revenue in the event the unit does not come back on line as scheduled, or in the event it does not perform as promised when it does come back, so the owner may insist on some form of damage recovery. However, there are several issues with this.

First, the contractor being asked to accept such penalties may not have a direct influence on the availability of the unit. Although the contractor may end up being late in completing this portion of the work, others may actually be the cause for the unit not being available. In this case, the owner could potentially collect liquidated damages from several contractors, simultaneously, although collectively the owner suffers less than the damages collected. To minimize this eventuality, the prudent bidder will insist on a "no harm, no foul" stipulation in the contract. This essentially says that if the delayed work by the contractor did not impact the loss of revenue of the client, then even though the LDs suggest payment, there will not be any, since there was no harm to the client due to the contractor's actions or inactions.

The second issue concerns pricing, which eventually affects the overall project cost. When bidders are asked to accept large liquidated damages, they will usually add additional monies to their bids to cover the unforeseen eventuality that they will have to pay some of these damages. If the project is to build a new power plant, the construction contract for this project could have a value of $25 million or more, with LDs of $25,000 per day. Although these LDs may then be capped at 20% of the contract value, the contractor is still exposed to as much as $5 million worth of penalties, a lot of cash. To mitigate the risk of losing this much money, the contractor will usually add contingency monies to offset some of these potential losses. To determine the amount to be added, one approach that is used follows next (fig. 2–8):

- Determine the number of days overrun to reach the LD cap.
- Estimate the number of days of overrun that *could* happen (in a worst-case scenario).
- Calculate what percentage the "could overrun" is of the "capped overrun."
- Take this percentage of the capped value as the contingency.

Contract Value	$25,000,000
Liquidated Damages Capped At	20%
Maximum Penalty Exposure	**$5,000,000**
Daily Liquidated Damages	$25,000
Maximum Days Exposure	**200**
Realistic Days of Overrun	30
Realistic Days/Maximum Days	**15%**
Contingency = 15% of $5,000,000	$750,000

Fig. 2–8. Liquidated damages contingency calculation

In this example, the contractor would add $750,000 to the bid, as a contingency to pay penalties for not completing on time. Per the calculations in figure 2–8, this equates to a 30-day overrun. Although this is only 3% of the total contract price, it is still three-quarters of a million dollars, the price differential by which many contracts are won and lost. And if one assumes that the contractor will not incur the penalty, and if the job is fixed price, this money will go to the contractor's bottom line, at the expense of the owner.

A more palatable approach to this issue is to move away from the punitive. Contracts that are structured with a win–win mentality will approach the liquidated damages somewhat differently. The first, and most straightforward, way is to include a payment of bonuses for early completion, assuming early completion has value, which is often the case for plant shutdown work. With new construction, there are often too many contractor interfaces to make early completion by any one contractor a value. However, when early completion does have value, offering a bonus may help offset the contractor's contingency. In the previous example, the contractor may look at the possibility of an early completion of 15 days and offset that against the potential 30-day overrun and now only add $375,000 to the bid, half of the original calculated contingency.

But there are also ways of structuring incentives to meet the end date that encourage both parties to work together to avoid delays. One such method is for the client to hold the LD contingency in escrow instead of paying it to the contractor as part of the base contract price. Then, if the contractor foresees a need for this money, perhaps to pay overtime to make up for schedule slippage, the client would release the money. The specific conditions under which this would be triggered would need to be spelled out in the contract.

The Prebid Meeting

So far in this chapter, we have discussed the basics of the bidding process, and we periodically referenced the prebid meeting. As the name implies, it is a meeting that is normally held before the potential bidders put forth a lot of effort to prepare their bids. Since some bid efforts can cost a bidder a million dollars over the course of the bidding effort and subsequent contract negotiations, it behooves each bidder to know (a) what the client is requesting and (b) who the competition is before committing to prepare a bid. The prebid meeting is the place for determining this.

The prebid meeting is generally held shortly after the project specifications have been issued and the potential bidders have had some time to digest the scope of work. It is normally designed so the contractor participants can ask questions of the owner or AE, and receive responses either on the spot or with follow-up correspondence. The participants in this meeting usually receive all of the information that is distributed at the meeting, as well as information that is provided after the meeting is over. The client's intent should be to bring all potential bidders to a level playing field. If one bidder has any questions about a particular issue, most likely others will as well. Therefore, the meeting gives each participant an opportunity to clarify issues.

But the prebid meeting is also a place for the potential bidders to size up their competition. Since attendance at this meeting is normally a mandatory requirement before being allowed to bid, usually there will not be any more bidders than the ones present at the meeting. This, then, affords the participants the opportunity to decide if their competitors have advantages that would make it difficult for them to be equally competitive. For example, if the project involves work at a plant where one or more of the attendees at this meeting are already performing other projects, these bidders may be able to exclude some of the mobilization costs that a newcomer could not exclude. The same also applies to demobilization, since this cost may already be covered in their current contract. A potential bidder may look at this and realize that this issue could not be overcome, and therefore decide to withdraw from the opportunity.

Another concern may arise when one of the potential bidders already has a special relationship with the owner or the GC. It could be that they both are part of the same corporate entity. That can weigh heavily in favor of that bidder. Also, there may be bidders who have vast amounts of experience in performing this work or working in the specific project location. A newcomer may feel outweighed by this. However, it could also be that these contractors are *not* at the prebid meeting. That should raise the question: What do they know (not to bid) that we do not?

The prebid meeting is also a time for the host, owner, and/or AE to size up the potential bidders. Especially if there are newcomers, the questions they ask or do not ask, could be indicators of their sophistication and experience for this type of project challenge. It is an opportunity for everyone to gain a bit more intrinsic knowledge about the project and those bidding for it.

Summary

It cannot be overemphasized that many of the problems that could be encountered during the execution of the project can be prevented during a well-structured bidding process. Therefore, properly structuring and managing this phase of the project development will go a long way toward ensuring a successful outcome. Making assumptions, without either verifying them or including them as part of the formal bid, can be disastrous. Communications is the key to preventing this. Since there are always at least two opposing parties in any bidding process, human nature pushes these parties into adversarial positions. This should be overcome, and an open and transparent dialogue established. The owner's perspective and the bidder's point of view should be considered by each, and misunderstandings corrected, before the bid is submitted. In the event that one or both parties do not have adequate expertise within their organizations, they should look to third-party support.

Once the bidding process is established, attention needs to be focused on the specifications. These documents can be voluminous, with hundreds of drawings and thousands of pages. Of course, not every potential bidder will ask to bid for the total scope of a 1,000-MW power station site, but all the bidders will have to understand how the piece of the work they are bidding on fits into the rest. The bidders should clearly understand the scope and clarify any unclear areas and interfaces before finalizing their bids. They should also understand the importance of the schedule; that no contractor is an island able to work in total isolation away from the other contractors. It must be clear that the schedule will have not only milestone dates to meet but also constraints that cannot be breached.

Then there is the all-important pricing and pricing-change mechanism. First, it is important that all parties to the project understand what the base price represents. Qualifiers may be used to avoid misunderstandings once the project is underway. But equally important is the method for changing the price as conditions on the project change. Since there will almost surely be changes in scope, conditions, or schedule on any large project, having a specific pricing-change mechanism in place will make this relatively straightforward.

Along with pricing goes cash flow or payments. Since the lifeblood of any contractor is cash flow, it behooves the bidders to carefully examine their proposed approach to the project. Developing a concise cash flow curve will readily highlight where shortages could occur. Since the successful bidder will be required to pay labor and suppliers in real time, the cash must be available to do so. If it is not, then the contractor will be forced to "buy" it from elsewhere, and this will increase costs, thereby reducing the profit margin. Unless this was planned at the outset (maybe the owner asked for some help in project funding), a cash shortage can be a very unpleasant surprise. Therefore, payment terms are a very important part of any construction contract.

Taxes are another area where, if not properly handled, contractors— and owners—can quickly find themselves in trouble. And this trouble can go beyond just the day-to-day activities at the site. This trouble can involve governmental taxing authorities imposing penalties and conducting in-depth audits that can last months beyond the completion date of the job. It behooves all parties to have tax experts involved during the bidding stage to ensure that everything is well planned.

Finally, an important part any major contract consists of penalties or bonuses. These usually start with the client seeking assurance that the successful bidder will adhere to the terms of the ultimate contract through some form of pressure tactic in the specifications. The most common form is the imposition of liquidated damages for failure to meet certain requirements, such as scheduled milestone dates. Most bidders will add some amount of contingency to their bids to help mitigate against this possibility. But by adding this contingency, the client now pays money to the contractor for an intangible, and it can be significant.

Many contracts do not offer an opportunity for earning a bonus. Were a specification to offer an opportunity for margin enhancement through bonuses, the typical bidder would reduce the contingency for LDs, affording the owner a job at a lower price. And then there is always a way to structure the contract such that the bidder puts no contingency into the bid—by having the owner keep the "normal" contingency amount in escrow, for use by the bidder in the event that issues arise that require extra costs to mitigate.

In the next chapter we move beyond the bidding "rules" to actually preparing the numbers to be submitted with the bid. However, throughout the process of preparing the numbers, the discussions in this current chapter should be heeded.

Preparing the Numbers   **3**

Estimating a construction project can be a daunting task, especially for the uninitiated. There are a host of factors to think about, such as man-hours, manpower, crew sizes, and shifts. There are the issues of labor rates, productivity, overtime, and travel subsistence. Escalation and inflation, personnel shortages, and equipment availability must all be considered. To add still more complexity, there is usually more than one way to perform (and therefore estimate) the work, and it is not always clear which way will be the most cost-effective, especially in the early stages of a project.

Although risk analysis and contingencies are often thought of as adders to the estimate, it is not always that simple. Does the estimate already include some inherent "fudge factors"? If so, too much doubling up on costs may cause the job to be lost to another bidder. Not enough, and the job could end up being a serious loser, or be canceled altogether.

Then there are the issues of job-site safety, quality, and environmental regulations. Has enough money been allocated to meet this project's specific requirements? Does this job require extra "hole watch" personnel, those workers that are not directly contributing to the progress of the work, but who must be there to ensure that the job is completed in a safe manner? Does the work require special safety training, clothing, or equipment? What about quality? Has enough money been planned into the estimate to ensure adequate resources for handling all of the hold-point inspections? What about keeping up with welder inspections and radiography? Or how about the massive amount of documentation required to prepare turnover packages when the job starts winding down? Were the necessary personnel for this considered when the estimate was being prepared? Have environmental regulations been investigated? Does the work affect ground drainage where spill protection or runoff must be controlled? Is it a "zero discharge" site? These lists can become lengthy.

But one thing is for sure: Once the work begins, there will be those who will find plenty of fault with the estimate and the estimator. They'll say things like:

- How could they have missed that?
- Did they even visit the site?
- They obviously used out-of-date labor and/or equipment rates!
- Didn't they see that there is no way to get this all done in the time frame allotted?

It never fails that the estimating group gets blamed for a host of irritants that frustrate the field crews as they try to execute the work and yet still stay within the budget.

This chapter is intended to help put a structure in place for preparing estimates that will minimize these kinds of issues. Topics that will be covered run the gamut from how to use proper estimating factors to crew sizing, labor sourcing, supervision, and proper tools and equipment that need to be integrated into the estimating process. There will be examples of how to format and structure an estimate. There will be a discussion of some of the various third-party sources available to assist in all of this. There will also be an introduction to differing ways to look at executing the work, with an eye toward planning a more efficient job that leads to a leaner estimate. But first, let's see how to get started.

Starting at the Beginning

What is an estimator expected to accomplish? In some cases, it is a straightforward assembling of unit rates to be used for a time and materials (T&M) project. At other times, it is a request to develop a cost-reimbursable structure that takes raw costs, adds markup percentages to cover the intangibles, and maybe puts a "not to exceed" cap on the total package. But then there are those projects that are intended to be turnkey, with a time-certain fixed price. These are the ones where the risk is the greatest, so the estimator is required to have an especially clear crystal ball for these. The estimator is expected to be able to predict the future, using today's data, and come up with a price that will ensure the project's completion, within budget, in the future. This is often a daunting and thankless task.

What does an estimator need to accomplish these tasks? First and foremost, the estimator needs an understanding of the work that is to be accomplished. If it is the construction of the 1,000-MW plant, in a remote location (as discussed earlier), the estimator has to be knowledgeable in a lot of areas. The estimator has to understand the engineered product or equipment that is to be assembled. This means knowing whether the tasks will require machinists and mechanics or riggers and welders. The estimator has to be able to envision where this equipment will be installed and what the conditions have to be before this can be done. Then, the estimator has to be able to determine what ancillary work is required to allow this to happen. For example, to estimate the cost of installing the gas and steam turbines in a 1,000-MW, 3 × 1 gas turbine plant, the estimator has to be able to envision the pedestals where this equipment will be placed. But further than that, the estimator has to be able to visualize the inter-connections between this equipment and the inputs and outputs that are required to make it function as designed. In other words, a good estimator has to become familiar with the functionality of the finished product and, oftentimes, with the engineering that went into designing and fabricating it as well. Certain parts of the turbines will have been shop-assembled and match-marked before shipment to site. The estimator has to know this and has to include time in the estimate for the precision work that will be required to ensure proper reassembly of these components in the field.

Proper knowledge of the end process of the plant will allow a good estimator to see the interfaces that will arise as the plant is being built. It will enable the estimator to account for work such as steam-line tie-ins, including the required preheating, welding, and stress relieving of the interface welds. It will also give the estimator the ability to include the time and cost of the final radiography, with an understanding that while this task is being performed, other work in the area will not be able to proceed, which affects the schedule and therefore the costs that must be included in the estimate.

An understanding of the ancillary work that will be required before the equipment can be installed is also very important. For example, knowing that (and how) a turbine pedestal has to be built will enable the estimator to factor in wait times and access restrictions. Since the pedestal can be a very large block of concrete, it takes time to place the forms, install the rebar, and pour the concrete. It also takes time for the concrete to cure, all being impediments to installing the T/G (turbine/generator) set. Once the pedestals are complete, access for setting the T/G sets may be more difficult, just due to their sheer size. This may create additional rigging needs, which then require that extra money be added to the estimate.

In addition to knowing how the equipment fits with the rest of the plant and knowing its physical relationship to other equipment in the same vicinity, the estimator also needs something else: tools that enable the estimator to assimilate the data of the physical plant, its location, and its ultimate functionality. Then, the estimator needs to be able to apply proven statistics to each work activity that will encompass the erection of this equipment. The estimator needs to have estimating factors that have stood the test of time, yet can be modified for the work at hand, if required. Finally, the estimator needs tools to be able to crunch numbers that eventually lead to calculating the bottom-line costs of performing the work.

The estimator needs to have access to data that can help to predict the future. Factors that will help predict economic inflation and escalation of costs (which are not necessarily codependent) will be required for proper calculation of the impact on today's cost in the future; it is part of the crystal ball clarity. The estimator needs to have a way to lock in costs that will be out of the control of the site staff once the project gets started. For example, performing radiography on a large supercritical boiler is costly. The estimator will want to obtain fixed, firm prices from radiography subcontractors for performing this work in the future when the job actually executes. The same applies to materials. It may behoove the estimator to establish fixed pricing for future deliveries of products such as concrete, rebar, and piping. Using cost predictors that are available in today's marketplace, applying them to projects that are proposed to be ongoing during the term of the job being estimated, and then weighing the probability of their accuracy, the estimator can negotiate for guarantees of price and delivery for materials and even heavy construction equipment. The estimator may have to be willing to pay some up-front fees to hold these prices or may have to be willing to place conditional orders with cancellation fees in the event the job fails to materialize. But these are tools that will allow the estimator to fix, or guarantee, the price.

The Site Visit

The estimator who does not visit the site where the work is to be accomplished starts with a handicap. The whole purpose of seeing the site firsthand is to gain a thorough understanding of the job-site conditions today and speculation of what they'll be like when the work actually executes. Information on items such as access to the site, access at the site, parking, and storage areas will be important for the preparation of the estimate. Contractor facilities such as subassembly areas, field offices, tool room availability, sanitary facilities, change rooms, and first aid must be investigated. Sources of utilities such as power, lights, water, and compressed air need to be identified. And finally, the local working conditions should be seen firsthand.

Not making a site visit is like trying to guess where the obstacles are in a walk across town, on a dark night, without any streetlights. Some obstacles will be large enough that they can be sensed or faintly seen. Others can be heard or smelled. But there are still many, many others that one would need to guess at—using an educational guess—such as where the curbs of a sidewalk are going to be. Preparing an estimate with only this type of information will lead to contingency money being added, which, as mentioned before, can either take the bidder out of the running, or if successful, place the job in jeopardy due to an insufficient cost estimate. This is why many estimators rely on another person to make the site visit if they, themselves, cannot go. But this can be fraught with errors.

Take the case of a project involving the expansion of a power generation facility in a remote location—Indonesia, to be exact. The job was to add an additional power-generating train—boiler, T/G set, and so forth. For an estimator from the bidding contractors to travel from the United States to this location would be expensive and time consuming. Therefore, one contractor decided to have his local representative visit the site. He was given a set of guidelines to use when making this visit. These guidelines touched upon most of the major issues described above:

- Site access
- Parking
- Storage area
- Subassembly areas
- Sources of utilities

Unfortunately, the representative decided not to visit the site himself. Instead, he telephoned the plant management and asked for answers to the questions in the guidelines. He did not go and personally look around. He did not see the labyrinth of overhead high-voltage lines, cable trays full of cables to operate the existing equipment, or the multitude of large- and small-bore pipe trains that crisscrossed the plant space, both above and below grade. Since he did not see any of this, and since the plant management did not spend much time answering the questions, these obstructions were not noted. To make matters worse, the vendor of the equipment was supplying the equipment in large, preassembled components to save on installation time in the field. This meant that large, heavy-construction equipment would be required to move the components on-site and lift them into position. The estimator, back home in the United States, therefore knew nothing about any of these conditions and proceeded to price the work in a fashion that did not include any efforts for temporary rerouting of these obstructions. Fortunately, at the very last minute, before the bid was to be submitted, the contractor found out that the representative had not visited the site. The contractor realized that his cost estimate for the work could be inadequate, and therefore he did not submit a bid.

There are other instances where things did not end so fortuitously. Take the case of the contractor who successfully bid to transport a preassembled boiler from the unloading harbor dock to the site, and then erect it in place within the plant. When the truck carrying the boiler from the harbor to the plant encountered a tunnel, it was no-go. The boiler was too large to fit through the tunnel. No one involved with estimating for this work took the time to actually travel the route that the equipment would have to traverse to get to the site; that is, no one made a thorough site visit. Unfortunately, there was no convenient bypass around this tunnel, so the boiler had to be brought to a side location, partially disassembled (i.e., cut apart), then the pieces taken through the tunnel to the site and reassembled. Quite a costly process. The contractor lost money, and the owner was not happy.

So how does one avoid such situations? The first approach is to listen to all of the horror stories such as these. Then, prepare a checklist, such as that in Appendix D to use for every site visit. Even this list is not complete. It will not guarantee that every abnormal job-site condition will be discovered. But it is a starting point.

Finally, there is the case of where a detailed site visit was made, including a personal trip from the harbor to the site. This diligence paid off through the discovery of a bridge that was not strong enough to sustain the load of a factory-assembled boiler. Therefore, the fabricator was instructed not to preassemble the boiler, and the estimator knew that the job would involve quite a bit more field work than if the unit arrived assembled. In the construction industry, this is known as equipment arriving in a knocked-down condition. The problem here was in the definition of knocked-down. This unit arrived as many, many disassembled pieces, far more than anyone envisioned. Why? Because the fabricator and the construction estimator did not communicate. Too many assumptions were made without verification. Although a site visit checklist would not have been of much value here, it should jog one's memory to ask still more questions, such as: "What does knocked-down mean"?

The Craft Labor Workforce

The craft labor workforce is the single largest cost component of any construction project. It is almost always at least 50% of the cost of the site work, and it can exceed 70% on major turnaround projects. It is also the single largest variable during the execution of the work; the labor can make or break the job, depending on how the workers are managed and how they respond to this management. But for purposes of preparing the budget, or estimating the actual work itself, two interrelated labor questions need to be addressed: (1) where will the labor come from (and will it be available)? and (2) what will be its productivity? Without going into the details of labor recruiting and productivity management (these are covered in more detail in chapter 11), we'll discuss how to tailor the estimating process using the answers to these questions.

We'll address labor being supplied under an original equipment manufacturer (OEM) contract. We'll talk about how the labor productivity of this type of arrangement differs from that of third-party labor suppliers or brokers. We'll also explore what to consider when preparing estimates based on the work being performed by in-house labor forces; productivity can suffer significantly under such an arrangement. We'll investigate the pros and cons of subcontracting the work outright. This approach depends on many factors, and they have to be taken into account when preparing the construction estimate, because in this instance the actual man-hours are usually the responsibility of the subcontractor, not the main contractor. And finally, we will address MWBE (minority and women-owned business enterprise) contracting. This arrangement also requires careful consideration by the person putting together the construction estimate.

Contracted OEM

The least risky method of contracting for the installation of major power plant components is for the owner or general contractor (GC) to put the responsibility of the work and performance of the equipment on the designer or supplier of that equipment. This is known as contracting to the OEM. Although there are many instances when this may not be cost-effective, from the perspective of the construction portion of the project, this type of contracting arrangement generally provides the longest term warranty that the equipment will perform as specified. In addition, if something goes wrong during the warranty period—for example, the forced draft fan suddenly starts to vibrate excessively—the OEM will usually fix the problem without recourse to the owner. The OEM would have the sole responsibility for a poor misalignment or rotor balancing job. If this portion of the work had been contracted to others, then a series of finger pointing would start, costs would increase, and time would be lost while responsibilities were determined.

When the project is designed for the OEM to install its own equipment, productivity frequently improves. The OEM generally has the best experience in installing this equipment. The OEM supervisors usually have been involved with the installation of this same equipment more than once, so they have seen where problems occur and how to work around those issues. Based on their experience and equipment familiarity, they can usually make more cost-effective work assignments than can those without this experience and familiarity. For example, let's take the installation of pulverizers for a 900-MW coal-fired boiler (fig. 3–1). The OEM's supervision generally has been involved with the installation of many similar pulverizers before. Additionally, most OEM supervisors periodically attend in-house seminars designed to increase their knowledge of the inner workings of their OEM equipment, such as these pulverizers. With this inherent knowledge, these supervisors can "see" the erection and assembly process from beginning to end more readily than supervisors without this familiarity and training. Additionally, they have better access to the manufacturer, the OEM, for the replacement of damaged or missing parts as well as access to the OEM's engineering department, which designed the equipment in the first place, at no additional cost to the owner.

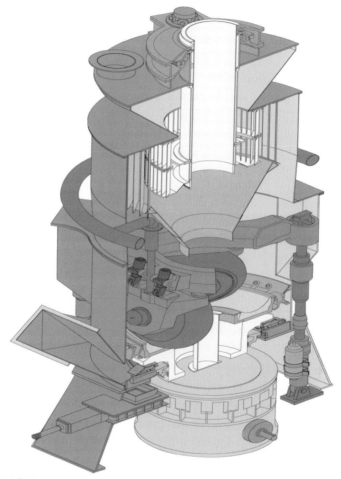

Fig. 3-1. Installation of specific equipment, such as this pulverizer, can most often be accomplished more effectively with the supervision of the OEM. (Courtesy of Riley Power Inc.)

An estimator for the owner or GC who knew that this equipment was to be installed by the OEM would not have to factor into the estimate any costs or downtime for potential problems such as missing or damaged parts or misfits due to engineering issues. Even if the installation work actually suffered additional costs or schedule delays, they would be to the account of the OEM, not the owner or GC.

Contracted non-OEM third party

When the higher cost of contracting to an OEM is perceived as not adding sufficient value to the project, the owner or GC may elect to contract to a third party. This would usually be at the expense of decoupling the warranty of the equipment from the warranty of the installation work. But when the equipment is not proprietary, or when there is no specialized equipment involved (e.g., civil work, structural steel, or electrical), then the OEM consideration is moot to begin with. In this case, the estimator would usually prepare a specification for the work scope in question and solicit quotations for it. The estimator would evaluate the bids, decide which one was most cost-effective for the scope in question, and include it as a line item in the estimate. (There is more on evaluating subcontractor bids later in this chapter).

However, when deciding to go to a third party to provide construction services, the estimator has to be aware of the positive *and* negative implications. On the positive side, third-party contractors frequently have experience working in the area where the work is to be executed. They are familiar with the labor and even may have "sweetheart" deals that allow them to hand-pick certain craftsmen for the work. They often have special arrangements with local tool and equipment suppliers because they work on various projects, often on a continual basis, in the same area, essentially being looked upon as a local employer and business merchant. Having other projects going on at the same time as the one at hand affords third-party contractors the opportunity to draw personnel, and tools and equipment from these other projects for the short-term needs of this project. In other words, the estimator would expect the site staff to have less concern about construction equipment and labor shortages with this type of arrangement, thereby not requiring a contingency for needing to import labor or construction equipment from other areas.

But there are downsides to using non-OEM contractors to install proprietary equipment that the estimator must also consider. In addition to not having easy, free access to information from the OEM, there may be the cost of OEM representation on-site, as required by the OEM to validate the work by the third-party contractor before the OEM will warrant the performance of the equipment. This can be costly because the OEM is usually the one determining how often and for how long this representative is required to be on-site. These costs need to be factored into the estimate. Other costs to be considered are the possibility of material damage in transit and repair to equipment damaged during installation. Without the OEM taking wrap-around responsibility for these types of issues, the owner or GC and the third-party sub may encounter lengthy delays and costs in trying to resolve them. This must be reflected in the project estimate.

Direct-hire labor

Whether it is the owner, the GC, or the non-OEM third party that is providing the labor, one of them is doing the actual hiring and firing. That organization's estimating personnel have the onerous task of taking responsibility for all the details, from wages to fringes, from burdens to taxes, from insurances to per diems, and from overheads to profits. Most if not all of these factors go into a formula that ultimately generates the labor rates to be used for estimating the project at hand. And all of these factors not only must be current but must also be escalated to the point in time when the job is anticipated to execute. Quite a formidable task.

To be able to determine what the wages and most of the other factors will be, the estimator will need to know from where the labor will be sourced. The first question about labor will be: "Union or non-union?" The answer to this may be dictated by the owner and plant. If the plant personnel are unionized, most likely the construction job will be union work as well. Or, the question could be answered when the contractor is selected. If the selected contractor is signatory to union agreements, the job also will be union. The next question then will be about availability (of the skills required). Will the locally available labor have the skills required to perform the work, in accordance with the productivity required to meet the project schedule? If not, from where can the labor be sourced, and what additional costs will this incur? When this situation arises, an additional labor source may be a third-party labor broker. Brokers frequently supply labor, either union or non-union, and make commitments for this supply at the estimating stage of a project. So let's look at these three potential labor sources, one at a time, and see how the decision to go one way or the other impacts the person preparing the construction estimate.

Union labor. When contractors use union labor, they operate under a legally binding agreement with each of the unions whose members they use, such as electricians, boilermakers, ironworkers, laborers, and pipefitters. This agreement essentially fixes the cost for the labor from each of the signatory unions for a defined period of time. The estimator can use these numbers with a fair amount of certainty that they will remain constant throughout the project. Sometimes, there will be adjustment factors, such as "a 2% increase in the base wage on March 1 of each year the agreement is in effect." If the agreement was signed for a three-year period, the estimator could easily calculate the labor cost for each year and apply it accordingly. Of course, if the project is expected to last longer than the labor agreements, then some crystal ball guessing may be required. Depending on the expectations of the workload after the expiration of these agreements, it may not be prudent to guess at the next set of labor contract conditions, and therefore it may behoove the estimator to use some caveat that ties the costs for that extended period of time to the terms of the future labor contract—a type of escalation clause. This will be discussed in more detail later in this chapter.

Irrespective of the labor contracts in place between the unions and the contractors, the owners of some major projects use the amount of the upcoming work as leverage to obtain special union rates. This is commonly called a PLA (project labor agreement). A PLA generally locks in the craft labor wages for the total duration of the project and of course obligates all project participants to use unionized labor. The persons preparing the labor estimates for work on a project should always ascertain if such an agreement exists, or may be developed, before finalizing the estimate. This will have no effect on the man-hours required to accomplish the work, but it can have a significant impact on the cost of those man-hours.

Non-union labor. Unlike the case of using unionized labor, when a contractor uses open-shop labor, there is no third-party organization with which the contractor signs long-term agreements that fix wages, fringe benefits, and other rules of engagement. Therefore, the contractor is free to pay whatever the labor market will bear and has the flexibility to be innovative with the benefits package: Sign-up and completion bonuses can be offered; productivity incentives are easier to implement; shift and working hour rules can be more flexible; and hiring, firing, and layoff requirements may be less stringent (e.g., often no last-in first-out require-ments for layoffs).

However, having more flexibility in the administration of the labor also creates more uncertainty in the final cost of these resources. Instead of being required to use an agreed-upon wage scale, the estimator must guess at the rates the site staff will pay once they hire the craftsmen for the project. On-site, the contractor's staff will frequently resort to manipulating the pay scale process to maximize workers' productivity, resulting in wage costs that are different from those that were used to estimate the job, which may lead some to question the competency of the estimator once job-site costs start being analyzed. But there also must be an offset. And there is. It is the opportunity to have lower labor costs since the union wage rates include more than just workers' wages and benefits, they also include the cost of the union infrastructure—its management personnel costs, the union office buildings, and so forth.

Labor broker. There are times when sourcing labor is not as simple as calling the union halls and requesting 50 certified welders, 15 mechanics, 6 electricians, and 20 laborers. The hall may not be able to supply them. Or non-union craftsmen may suddenly not be available, locally, in the quantities the job will need. So the contractor may turn to a third source of labor, a labor broker. A labor broker is an organization that has a list of potentially available people that supposedly meet the skill requirements of the job. In fact, some of the larger brokers will even pretest personnel in the presence of a contractor representative at no additional cost, before sending them to the job site. These organizations usually require a signed contract with the contractor, fixing the hourly or daily rates and fringes for the duration of the project. When other ways of sourcing labor are no longer as assured, the estimator may want to consider pricing the work using this option.

Sometimes, using a labor broker is more cost-effective than the standard union or non-union approach. Pricing craft labor from a labor broker can result in lower overall labor costs, although not always immediately apparent. For example, labor brokers usually handle all of the administrative details of paying the workers and submitting the governmentally required taxes, insurances, and so forth, to the appropriate government entities. They also will handle travel, and sometimes living arrangements, take care of per diem payments and even make sure the employee's check is deposited wherever the employee requests. In other words, labor brokers frequently will provide all of the administrative services that otherwise would be the responsibility of the employing contractor. This, therefore, relieves the contractor of the necessity of having a payroll staff on-site—at least for these workers. It also may reduce other on-site administrative staff that would normally assist in the management of the hourly craftsmen. Although not directly reflected in the labor rates of the individual workers, these avoidable costs will be reflected in the overall job-site cost reduction.

In-house labor

There are some projects, especially on turnaround work, where a portion of the scope will be performed with labor already on the payroll of the plant owner. This labor may be union, or it may be open shop. Situations for using in-house labor generally arise when the plant management decides to self-perform some of the work. It may also arise when the plant management asks the contractors to use the plant personnel to keep them gainfully employed, especially when a unit is off line during an outage. Some plant management personnel also feel that by using craftsmen from within the plant labor ranks, efficiency is enhanced since the workers are already familiar with the plant and its working rules and requirements, and the workers have a loyalty to the plant.

When contractors are asked to use the local plant labor, they must be careful when preparing the project cost estimate. As well-intentioned as the plan may be, if the work takes place in or near an operating plant, and if an emergency arises at the operating plant that requires plant-specific expertise to repair, these workers often will be pulled away from the contractor for whom they are working to attend to the emergency. This will affect productivity, possibly schedule, and definitely project cost. The estimator for work in this instance must be very specific about the assumptions made when agreeing to such an arrangement. For example, the estimator may require that the portion of the work the plant personnel perform will not be work on the critical path of the project. The estimator may go further and require that in the event other workers have to be brought from elsewhere to replace the removed plant workers, any additional costs will be reimbursed.

Subcontracting

There are many reasons to consider subcontracting parts of the work. The reason can be a previous, favorable affiliation with a particular subcontractor. It can be to keep the work within the corporate family of contractors, or it can be as simple as the economics of who can do that scope of the work for the lowest cost. But when initially laying out the workflow plan, subcontracting can be a viable way to shift the responsibility of resourcing labor, and tools and equipment, while simultaneously mitigating risk.

It is not uncommon for the owner or the engineering, procurement, and construction (EPC) contractor to subcontract specialty tasks such as civil, electrical, and insulation work. Also, specialty machining and even specialty welding is often subcontracted since the skills required by these disciplines are often not available from the sources of labor that the owner or EPC contractor normally use. But actually locating and contracting for these services can be a challenge. Issues such as union versus non-union, differing pay scales, distance from the job site, and the per diem for travelers all enter into the decision-making and estimating process.

Subcontracting also brings with it a whole host of issues that are nonexistent when self-performing. Similar to the effect that multiple projects in the area have on the supply of labor, subcontractors also can get stretched too thin. Some will accept the work and then not be able to perform, while others will accept the work, do an unsatisfactory job, and ultimately cost the project more time and money than if the work were self-performed.

It is also very important to plan for the administration and management of the subcontractors at the bid stage. Since misunderstandings usually occur due to poorly planned processes, issues such as claims can be greatly minimized if properly addressed before the estimates are finalized.

Once the decision has been made to subcontract part or all of the work, the issue raised next will be: to whom? There are various ways to determine this. As mentioned earlier, it may be to use a corporate family contractor, thereby keeping the work in-house. It may be based on previous experience with a particular contractor. It could be based on contractor reputation or recommendation by others. However it is determined which contractors will be invited to bid for the work, there are some basic fundamentals that should apply to them all. First and foremost is ability. Can the contractor obtain the resources required to perform the work, within the project budget and schedule constraints? Then, can the contractor do so safely and within the confines of the quality requirements set out for the work? The answer to the ability question is generally straightforward, and one of the best ways to verify this is to visit the contractor's base of operations as well as several ongoing projects—a "see for yourself" approach. However, the answer to the safety and quality question is not so apparent.

Subcontractor safety. Let's explore safety first. Over the past 40 to 50 years, the industrial contracting industry has gone from perceiving safety management as a nuisance to managing safety performance as a business in and of itself. Public opinion, governmental regulations, insurance costs, and worker attitudes have refocused management attention. In today's contracting world, there are few, if any, industrial contractors that do not have very formalized programs in place to manage job-site safety and provide evidence that they are doing so successfully. When deciding which subcontractors meet the safety requirements for the work for which they will be asked to bid, there are very straightforward, logical processes that can be used. See chapter 9 for more on this.

Subcontractor quality. Contrary to the historical data and detailed reporting requirements normally found in a contractor's safety program, quality program results are frequently less formalized and therefore less available to GCs trying to use information to prequalify subcontractors. However, this should not deter one from asking the potential subcontractors for a demonstration of sound quality policies *and* results. At a minimum, the proposed subcontractors should provide a copy of their company quality policies and an uncontrolled copy of their quality assurance (QA) manual. It would also be helpful to require a copy of a job-specific quality plan from a project similar to the one being contemplated, as well as typical work instructions and sample turnover packages. From here on, statistics can be requested; however, they may be difficult to come by and even more difficult to interpret.

For example, one can ask the bidders to provide examples of cost-of-poor-quality analysis from previous projects along with a tracking system that (ideally) shows consistent improvement. Other requests could include proof of ISO (International Organization for Standardization) certification, proof of an inspection and audit analysis, corrected nonconformance issues, proof of ongoing personnel training, and also verification of a process that leads to vendor approvals. See chapter 8 for more on this.

Subcontractor evaluation

Subcontractor vetting and evaluation should be one of the pillars of project risk-management processes. But is it really done with sufficient rigor? Is it a formalized process that defines up-front requirements as they potentially impact project risk or failure? Is it done in a manner that relates to the job at hand and not just based on generics, past history, or luck? Does the vetting process force attention on how the bidders plan to approach the specific job at hand, how they plan to provide the resources for that particular job, and who they will assign to be a part of that specific project team? This should be a part of the vetting and evaluation process because price should not be the only criteria.

Two important aspects of subcontractor vetting have just been discussed, safety and quality. Briefly mentioned as well was how references and firsthand visits to active project sites can go a long way toward determining a sub's ability. But what about determining if the subcontractor *really* understands the work that will be performed? What methods can be used for making this judgment?

One very basic method is the weighted value approach (discussed in more detail in chapter 10). In simple terms, the weighted value approach means that the work scope is broken into a discrete and logical number of activities. For example, if the work involves a lot of boiler, turbine, and condenser work, one could make a list of the activities to be required as follows:

1. Mobilization
2. Boiler superheater work
3. Boiler burner throat or cyclone repairs
4. Turbine reblading
5. Condenser retubing
6. Condenser waterbox rebuild
7. Other work
8. Demobilization

This is a list of eight activities that one could expect the subcontractor to undertake for this repair work project. As a separate exercise, the estimator could then apply his or her own estimate of man-hours to each activity, add up the total, and determine the weighted value of each activity as it relates to the total work. This could be considered the baseline.

Next, the person preparing the subcontractor specifications would ask that each bidder subdivide his or her bid into the above categories and provide the expected man-hours accordingly. After the bids were received, the evaluator would first compare the weighted values of the bidders' categories to the evaluator's own list. If the comparative values were proportionately equal to what the evaluator had, then at a minimum, one would expect that the bidders understood what was being asked of them. If the bidders' values were not in line with the evaluator's, the first question to be asked would be if the bidders understood the work. Then, the bidders' weighted values should be compared to each other. This would enable the evaluator to see which bidders were in line with each other, and which were not.

Figure 3–2a shows a compilation of bids received from various subcontractors ABC, DEF, and so forth. The bids are arranged such that they can be compared to the estimator's baseline bid as well as to each other. An interesting fact emerges immediately—the average number of man-hours across all four bidders is almost exactly what the estimator thought would be required to perform this work. Although not shown in the table, one can also eliminate the high and low bids and average the remaining bids. The result? Also almost exactly what the estimator had predicted. This goes a long way toward validating the estimator's baseline numbers.

In this first tabulation we see some interesting pictures. Bidder ABC has the same man-hours as the estimator has for the total superheater and cyclone work, but has them reversed. Does that mean that ABC made an error when filling out the bid form, or does it mean that ABC does not understand the work scope? Also, ABC's man-hours for the turbine work seems extremely high, while that for the condenser work looks very low. And, the total man-hours seem low compared to all of the other bidders. We can make similar analyses with the data provided by the other bidders and reach some preliminary conclusions.

But let's go a step further and do some real evaluation. Figure 3–2b shows the same data as above, but in a weighted value format. The value of this information is that one can make a quick evaluation of which bidder comes the closest to understanding the work. By comparing these weighted values to those of the baseline, one can pick out which bidder comes the closest to the baseline in the most categories. And the winner in this example is? Bidder DEF, who comes closest in seven out of the eight categories. The only category of concern is the turbine work, and that could be discussed for clarification.

Activity	Baseline mhrs		ABC mhrs	DEF mhrs	GHI mhrs	JKL mhrs
1. Mobilize	2,500	2%	3,000	2,500	3,000	3,000
2. Boiler superheater work	40,000	27%	25,000	45,000	65,000	20,000
3. Boiler cyclone repairs	20,000	13%	35,000	25,000	30,000	40,000
4. Turbine reblading	15,000	10%	25,000	10,000	20,000	10,000
5. Condenser retubing	40,000	27%	25,000	40,000	10,000	15,000
6. Condenser waterbox rebuild	20,000	13%	10,000	25,000	15,000	30,000
7. Other work	10,000	7%	6,000	8,000	20,000	20,000
8. Demobilize	2,500	2%	3,000	2,500	3,000	3,000
	150,000	100%	132,000	158,000	166,000	141,000
Contractor average man-hours	149,250					

Fig. 3–2a. Subcontractor bid man-hours

Activity	Baseline mhrs		ABC wt val	DEF wt val	GHI wt val	JKL wt val
1. Mobilize	2,500	2%	2%	2%	2%	2%
2. Boiler superheater work	40,000	27%	19%	28%	39%	14%
3. Boiler cyclone repairs	20,000	13%	27%	16%	18%	28%
4. Turbine reblading	15,000	10%	19%	6%	12%	7%
5. Condenser retubing	40,000	27%	19%	25%	6%	11%
6. Condenser waterbox rebuild	20,000	13%	8%	16%	9%	21%
7. Other work	10,000	7%	5%	5%	12%	14%
8. Demobilize	2,500	2%	2%	2%	2%	2%
	150,000	100%	100%	100%	100%	100%

Fig. 3–2b. Subcontractor bid evaluation

Minority and women-owned business enterprise (MWBE)

When a project attracts the attention of the public, and when the public is less than enthusiastic about it, the project often suffers, especially in the area of productivity. One way to mitigate this is to involve the public in the project by planning to use some of the locally owned MWBEs. (This may also be a requirement from the owner, or an internal corporate business requirement). However, using MWBEs requires some thought and preplanning during the estimating stage of the work. Quite often, these enterprises are small. They may be inexperienced in the power plant construction world. They may not have a stellar safety record. They may not have a formalized quality program that has been audited by a third party. They may not be ISO certified. But on the other hand, they may have the political connections within the community to help reduce the "roar of disapproval" that some power plant projects attract. For this reason, the estimator may need to find ways to include some of these MWBEs in the project.

The first step will be to clearly identify, and obtain agreement, on areas requiring special support. For example, a local, woman-owned electrical contractor might be an ideal candidate to use due to the owner's connections with the city council. Although this contractor may be completely inexperienced in the heavy industrial construction world, she could, most likely, handle peripheral duties such as the installation and maintenance of temporary construction lighting. She probably could also install the project's permanent lighting system, if this was part of the overall work scope. And she most probably would want an opportunity for her company to learn about the heavier work such as elevated cable tray installations, high-voltage cabling, and motor control center work.

But to arrange for her to be eligible to bid for some of this work, certain exceptions will need to be made to the standard requirements that most bidders must meet. For example, the experience requirement may have to be waived. In place of the waived requirements, a plan of supplementing this contractor would need to be developed. The estimator, after determining where support is required, should add to the estimate the cost for this additional support.

For example, this contractor may not have a formalized quality program normally expected for work on a project of this magnitude. To offset this shortcoming, an agreement could be reached whereby the MWBE would work under the auspices of the main contractor's quality program and would be trained on the job. This would be similar to the MWBE acting as a labor broker with respect to following the quality processes and procedures. The person putting the job estimate together would then include the cost of an extra part- or full-time quality person who would be providing this support.

The Supervision

Once the labor requirements of the job have been established and estimated, the next step is to plan the supervision required to manage the labor. If the work is completely subcontracted, the supervision required will be much less than if it were all direct-hired. For example, if an owner hires one major GC for all of the work, this owner, when preparing the project budget, will only add money for staff sufficient to oversee the project basics. Most often, there will be personnel to oversee the following:

- Safety
- Quality
- Schedule
- Engineering
- Administration
- Community relations

The day-to-day labor management will be the responsibility of the GC, and the staff to do this will be included in the pricing.

The GC, in turn, will most likely subcontract parts of the work. Referring back to the 1,000-MW gas-fired plant, the work the GC frequently subcontracts is the heavy mechanical (boiler, T/G, and balance of plant), the electrical, the specialized piping, and the chimneys. The basic site works and even the major civil work may be retained by the GC. Therefore, the staff that the GC's estimator would include would be a mixture of subcontractor administrators and direct labor managers. In addition, the estimator would include personnel to interface with the owner. Also on the GC's staff would be personnel for managing the site safety program and the project quality program. However, the GC also frequently needs access to engineering staff for equipment interface issues and especially for developing or reviewing heavy rigging plans. This support staff also needs to be added to the cost estimate of the job.

Once the individual staff needs have been identified and their durations determined, their costs must be developed. If they come from within the organization, there will be a costing rate that includes the salaries, benefits and burdens, and any assigned overheads that the accounting system assigns. This value is normally what is used to calculate the costs of the in-house personnel. Then, their cost of living has to be decided. Will they live away from home, or will they move their home to the job? Each has its own cost benefits. However, contrary to the old ways of building power plants, where many supervisors traveled around the country and lived in mobile trailers, today's supervisor usually commutes to and from the job site every few weeks. So the cost of living for a supervisor today is generally a function of a hotel or apartment near the job site, fully furnished with utilities included, a car, and airplane tickets back and forth to the supervisor's home. On top of that, a per diem may be provided for food and other incidentals. All of this must be included in the estimated budget.

If the supervision is sourced from outside of the company, there are generally two sources: (1) rank-and-file craftsmen and (2) third-party staffing agencies. Each has benefits as well as drawbacks.

Rank-and-file craftsmen

Supervision sourced from rank-and-file craftsmen is frequently referred to as craft supervisors. They normally live in the area and are paid slightly above the foreman pay rate of their trade. If the job is a union job, the union fringes must be added to their cost, the same as with any other union craftsman cost rate. The advantage of using supervision from within the ranks of the workforce is that they usually have an intimate knowledge of the personnel on the job. They will know the strengths and weaknesses of most of the workers who come from their union local because they will have worked with them on previous jobs. The disadvantage is that they will often be more loyal to the needs of the union instead of the contractor.

If the decision is made to use supervision from the crafts, there may be support costs in addition to the loaded rate for their time. If so, these costs need to be included in the estimate. For example, do they need a vehicle, or will they already have one since they probably live in the local area? Will they need a per diem to offset meals and other incidentals, or will they be expected to provide their own meals? Not accounting for these costs at the estimating stage can skew the site operating budget.

Third-party staffing agency

When the decision is made to use supervision from outside the company staff and outside of the local labor pool, the supervision is frequently sourced through third-party staffing agencies. This supervision is often referred to as contract hire supervision due to the nature of the assignment. Different from craft supervisors, who usually return to the rank-and-file workforce upon completion of their assignment, contract hire supervisors normally go from job to job as assignments materialize. As Rick Sparra, president of Defined Source Cooperative, one of the industry's leaders in providing these support services, said:

> The continuing uncertainty of build, do not build, perform maintenance, do not perform maintenance, the roller coaster decision making environment the power generation industry finds itself in today plays havoc with planning for staffing of construction projects. This applies to owners as well as contractors and that's where third-party staffing agencies like ours can be invaluable.
>
> We have access to supervisory personnel, many of whom used to be employed by these owners and contractors. Therefore, they have a broad range of experiences in the building, and rebuilding, of power plants. Some of these people now are semi-permanent employees of ours. They may be covered under our insurance plans, both health and life. They may have retirement benefits. This recreates a loyalty to us similar to what was seen when they were employees of those owners and contractors.
>
> We, similar to some of our peer competitors, provide these personnel on a defined term/contract basis. Our clients are able to access and utilize all levels of industry talent whenever needs arise, and for only as long as they require, while still realizing imperative and successful results on their projects. Most of our clients consider us as part of their teams and a necessary extension of their resources, not simply a "vendor."

Due to the lack of major power plant construction activities throughout the 1990s, and even many of the years since then, there are not enough experienced construction supervisors to handle the workload anticipated over the next 15 to 20 years, even without new-build, coal-fired plants. And if the potential for a renaissance of nuclear power plant construction is included, the need for more personnel increases exponentially. This is where these third-party staffing agencies can be invaluable. However, there is a cost.

Generally, the rate charged by these agencies includes recovery of the cost they incurred to locate the specific individuals. These costs will also include a profit margin and sometimes additional charges, especially if the agency is being asked to absorb the cost for payroll, transportation, per diem, and the like. Also, if the contracting entity decides it would like to hire the supervisor on its permanent staff, there is usually an additional fee, maybe up to half a year's salary, that could be incurred. But in today's environment of personnel shortages, using third-party agencies is becoming more and more a part of normal business.

Tools, Equipment, and Materials

Once the estimating process has proceeded through the labor and supervision stage, the next focus must be on the tools and equipment that the labor and supervision need to perform their tasks. At this point of the process, the job will have been segregated into many, many small steps, or work breakdown structures (WBSs), either by the estimator or by others on behalf of the estimator. The estimator will then have to review each WBS to determine what types of tools and equipment will be required to perform the task within the man-hour estimate allocated for that task. The estimator will have to look at each WBS from the point of view of the superintendent and craftsman actually performing the work. In addition to the costs of tools and equipment, the costs of hand tools and consumable materials need to be included. These may seem inconsequential at first, but they can become a major cost component of the project, so they should be included either as a percentage of the labor cost or as a specific dollar amount per man-hour. We'll examine each category.

Also important, yet sometimes overlooked during the estimating stage, is the cost of shipping the tools, equipment, and materials. Heavy equipment transportation costs can be very significant, especially if special road escorts and permits are required. Smaller equipment can frequently be shipped in containers, or just crated or strapped onto pallets. However, small equipment will still require special-handling gear like forklift trucks or small hydraulic cranes, both at the shipper's facilities and at the job site where the equipment will be received. The cost of this, plus the freight and freight insurance costs, should be included in the estimate. The same applies to any project or construction materials being shipped, especially if some of it requires air freight.

Heavy equipment

The most obvious tools and equipment component that needs to be considered is the heavy equipment: cranes, bulldozers, graders, dump trucks, and so forth. For an initial greenfield project, a large quantity of earth will probably need to be excavated, moved, and leveled. Large earthmoving equipment, sometimes GPS controlled, will be needed. The estimator will have to include the cost of this equipment in the project.

While reviewing the estimate and the proposed work plans, the estimator should constantly be on the lookout for opportunities to reduce costs. For example, if one of the tasks is to install air heater baskets, most likely the estimator will have planned for this to be done either using a tugger arrangement or some kind of crane. Now is the time to review which is the most cost-effective approach—tugger or crane? If a tugger, its costs—installation, rental, and operation—must be added to the estimate. If a crane, then its costs must now be added to the estimate—and in addition to rental and operation, there may be major transportation, setup, and breakdown costs that need to be included, as discussed above.

If the contractor owns the equipment, usually there will be an internal costing rate, dependent on the age of the equipment that will be used. If the equipment has already been depreciated, the cost may be very low. If it is new or not yet purchased, the costing rate may be quite high, depending on how many years it will be depreciated. The estimator must be aware of all these factors, so the costs built into the estimate will reflect the costs that the project actually incurs once underway.

If the contractor does not own the equipment, then the equipment will most likely be rented from a third party. This is where timing becomes critical. As with everything else in this industry, the availability of heavy industrial construction equipment is also volatile. Just a few years ago, the North American construction upswing was expected to last for quite some time, with heavy spending not only for power generation facilities but also for highways and mass transit. It was thought that over the following decade, crane rental firms would have to purchase more equipment just to replace thousands of crawler cranes that were more than 25 years old. Adding that to the strong outlook, at that time, in emerging countries such as Brazil, Russia, India, and China, with 20% growth each, there seemed to be a serious shortage of cranes to meet those growth targets.

How times have changed! As of the time this book was being written, this shortage has not materialized. However, that is not to say it will not happen in the future. Since many power plant construction projects are estimated, and even bid, two or more years before the work actually begins, thought must be given to how equipment availability can be guaranteed for years into the future.

So what is a prudent estimator to do? During the estimating phase of a project, the prudent estimator works with the heavy equipment suppliers to lock in the supply and price, or to include terms that pass on price increases, delays, and the cost of delays to the hiring contractor, GC, or owner.

There is another way, too. Although generally not under the purview of the estimator, special arrangements can be made between the contractor (or owner) and the crane supplier. Alliances can be formed whereby the equipment supplier and the contractor agree on a long-term relationship. Basically, the supplier guarantees availability and price, and the contractor agrees to use only this supplier's equipment.

Finally, as existing cranes have aged, so have their operators. But without many new cranes coming on the market, additional operators have not been brought into the industry. The prudent estimator, when preparing the cost for cranes, should also include the cost for personnel to operate these cranes. This is especially important because some of the newer cranes are different from their predecessors. The operators may need specialized training, and this must be figured into the job cost estimate.

Small tools and consumables

After wrestling with the price development of the heavy equipment needed to build the power plant, the estimator has to look at the small tools and the consumables that will be needed by the craftsmen on-site. Small tools are usually considered those tools that the accounting department allows the site personnel to expense upon purchase at 100% cost to the job. Tools like this are generally considered hand tools that normally would not be repaired in the event of breakage or reused on a future project. For example, hammers, chisels, wrenches, cutters, and reamers fall into this category.

Job-site consumables also fall into this category. Rags, oils, greases, grinding wheels, and any other items that are not reusable are normally considered consumables. Just like small tools, they would be expensed as soon as they are purchased, whether used or not, and disposed of at the end of the job.

The estimator often has a difficult time deciding how much to allocate for these items. Sometimes, it is just a guess. At other times, it is based on records from previous projects. For a major power plant construction project, these items can be equal to as much as 20% of the labor cost. However, when the labor rates—and hence costs—are very low or very high compared to the records on which the 20% number was derived, it may be better to use a number of dollars per man-hour. This is where a good estimator can be worth one's weight in gold. Underestimate such a 20% item, and the job could be a loser from day one. Overestimate the same item, and the job may never be won. So it behooves the estimator to be very clear on which basis these small tools and consumables are estimated.

Materials

Finally, let's look at what should be included in a construction estimate to cover the cost of materials. The first step will be to define materials. Ultimately, it has to do with the manner in which these items are accounted for by the project's accountants. If the materials form part of the permanent plant, quite often they must be capitalized so they can be depreciated along with the rest of the plant equipment. If this is the case, they cannot be included in the construction estimate in the same manner as the labor, supervision, tools, and equipment. In fact, it is usually more advantageous to exclude materials that form part of the permanent structure from the construction estimate. If the construction contractor is required to furnish any of these, it is often more advantageous to keep their costs in a separate budget (and estimate). In this manner, their ownership and costs can be more readily transferred to the plant without the chance of mixing capitalized and noncapitalized items.

But what about those materials that can be expensed? Here we are talking about items such as temporary support steel, fit-up bolts for the main steel structure that will be replaced with the permanent bolts, scaffold boards, and a host of other "temporary" items. These exact a cost on the project and therefore must be included in the job cost estimate. As with the heavy equipment, the estimator needs to review each WBS work activity and determine what temporary materials might be required. For example, if the project was the replacement of superheater elements in a large boiler cavity, the work plan may be to erect a temporary monorail structure from where the elements are planned to be extracted from the unit to the point where they will be lowered to the ground (and the new elements raised and inserted into the unit in a reverse manner). This may require special steel for the monorail and special steel to be attached to the boiler support structure to hold up the monorail. The cost (and availability) of this material would have to be included in the estimate.

Another example is an on-site prefabrication facility. Suppose that the plan for building a new boiler included preassembling the waterwall panels, along with their seal boxes and buckstays, or preassembling superheater elements. To do a good job would require a series of prefabrication tables, set up in the preassembly yard (fig. 3–3). If the boiler was large, there would be a large number of tables, each with the ability to support many tons. This would require many sheets of flat steel plate, many tons of structural beams to support the flat plate table tops, and possibly some concrete and rebar to form the foundations upon which these tables would rest. The cost of this material and the cost of fabricating the tables and setting up the preassembly area would have to be included in the estimate as well.

Fig. 3–3. Preassembling boiler waterwalls (Courtesy of Construction Business Associates LLC)

However, for some projects there are times when prefabrication or preassembly work is done at the manufacturing facility, for example, heat recovery steam generators (HRSGs). Since these boilers are not as huge as a typical coal-fired unit, they are sometimes almost completely assembled at the manufacturer's location and then shipped to the job site by special heavy-haul equipment (see fig. 1–1). In this case, the construction estimator need not be concerned with prefabrication time and costs.

There is one other task that must be undertaken: ensuring that these large units can be safely routed to the intended site. Earlier in this chapter, "The Site Visit" section described one project where a boiler was assembled at the manufacturer's facility and barged to a port near the site. But when the unit was being transported from the port to the project site, on a highway that went through a tunnel, the load would not fit. It had to be towed to another location, cut apart, transported to the site in pieces, and then reassembled. No estimator had included the cost or time for this.

Quality Control in the Estimating Phase

Normally, one would not expect to spend much time estimating quality control costs. Generally, a construction estimate is thought of as labor based, supervision wrapped, and tool and equipment supported. If additional costs are warranted, they would be added as a percentage of the bottom line. But this is just not realistic. Just as was addressed during the subcontractor evaluation discussions earlier in this chapter, quality control and safety are two areas that are usually given inadequate attention during the estimating stage of a construction project. Here, we'll address quality control, and we'll address safety in the next section.

Quality can never be overemphasized. It must be inherent in a project. But without a defined process, it can easily take a back seat to the importance of completing the work. Therefore, it is incumbent on the estimator to specifically address it. When the site gears up for starting the project, its management needs guidelines to follow for organizing and staffing the quality department. The first place for them to look is in the estimate for the project. What was included? How many people? For how long? Then, when they plan for the work that has to be done by the quality team, they will have an idea of how much money was included for this. If the estimator did not include a line item for this, instead only assuming a percentage of the bottom line, the site management has little idea of where to begin.

Good estimators will look at the site quality group the same way they look at labor supervisors. They will envision a group headed up by a leader, complemented by persons skilled in the disciplines of the work that will be encountered. For example, if the work starts with site clearing, excavation, subsurface structures, and grading, then there should be someone planned into the work to oversee the quality process; someone to ensure that hold-points are honored and tests performed (tests such as soil compactness, concrete slump, and concrete core samples). This function will certainly come to an end long before the main mechanical work of the project commences, so the estimator would only include these persons for that duration, but they would be included there, along with all of their relevant salaries, fringes, cost of living, and office support.

As the project work continues, these skill needs would change from civil type work to mechanical and electrical, among others. The person preparing the estimate for the quality function would then estimate the cost for personnel to oversee the quality of the main plant erection, including structural steel alignment, pressure part welding, electrical systems installation (continuity checks, ring-outs, meggering, etc.), and all of the other details that will be required to assure the completion of the plant construction in accordance with the contracts, codes, and specifications. Again, the cost of these peoples' salaries, benefits, cost of living, and support services would have to be included. No small feat.

And one of the areas almost always underestimated is the preparation of the turnover packages. As the construction work progresses, end points are reached that require turnover from the constructor to the start-up group. Most project specifications require that the constructor prepare a set of documents that verify, with formal signatures, that specific areas or pieces of equipment have been installed in accordance with requirements and that they have been tested accordingly. Let's take the example of a boiler feedwater pump. To enable the constructor to turn this pump over to the start-up group, who will test and certify that it is ready for service, the constructor will be required to provide documentation that verifies that this pump is ready for turnover. This package of documentation will have to include verification that the pump is properly wired and has been bumped; that is, when the button is pushed to turn it on, it will rotate in the correct direction, at the correct number of revolutions per minute (rpm), draw the correct amperage, and not overheat. Additional verification will be required in the turnover package for proper alignment and balance. In other words, there must be paperwork, signed by the constructor and the start-up personnel that the pump has run, that it did not vibrate and that it ran true.

Preparing these types of turnover packages requires a lot of effort on the part of many people. It is usually relegated to the constructor's site quality personnel to make this happen. Therefore, the cost of these people must be included in the job cost estimate. This cannot be a bottom line percentage adder to the cost of the job.

Safety in the Estimating Phase

Like quality, safety management is also an integral part of every construction project, especially a large power plant construction project. Again, instead of allowing it to be an arbitrary percentage of the bottom line of the estimate, safety should be estimated as a separate identity in the cost of the job. On large projects, one safety person is seldom sufficient to ensure that the workers have a safe working environment. Some organizations require at least one dedicated safety officer for every 75 craftsmen. Others take a much more stringent approach by requiring that any project work shift having over 25 workers and a projected duration of two weeks or more must have at least one full-time safety officer. With workers' compensation insurance costs averaging somewhere in the neighborhood of $7.00 per man-hour worked on major power plant construction projects, it is only prudent that the job estimate be built up to include adequate safety supervision, tools, and equipment.[1] We are talking over $1 million savings on a large project if properly managed!

In addition to providing costs for the safety supervisor—some still like to call that person the "traffic cop"—the cost impact of the others involved in safety must also be included. For example, many projects involve craftsmen working in confined areas. This type of work requires a second person, commonly called a "hole watch" person, whose sole function is to keep an eye on the workers inside the confined area. In the event an emergency arises, this person is available to assist in addressing the emergency and quite possibly saving the lives of those inside. However, this person is not actively contributing to the productivity of the work and therefore is often not included in the man-hour estimate of the project. Because this position is a cost to the job, it must be included somewhere. The appropriate place is in the section of the estimate reserved for the costs allocated to safety.

Other safety functions frequently overlooked during the job estimate are inspections of the excavations for and pouring of concrete work. If the excavations are improperly shored, they can collapse, trapping and sometimes killing workers. But like the hole watch person, the individuals inspecting shoring and monitoring large concrete pours are not directly adding productive hours to the job progress. They are trying to prevent accidents, which would take away from the productivity and the bottom line of the job.

In addition to on-site safety manpower, there are also on-site safety equipment costs that should be included in any job estimate. Personal protective equipment (PPE) is usually required for every person on-site, whether they are direct labor craftsmen, indirect labor yard personnel, or supervisors. This equipment ranges from safety glasses to hard hats to safety belts to reflective vests and beyond. There is a cost for this, and it should be included in the job cost estimate. One easy way to do this is to allocate a number of dollars for each person identified in the personnel loading chart for the project, including indirect and supervisory personnel. For high-visibility projects, additional PPE may be needed for the continuous barrage of visiting management and dignitaries.

Most major construction projects also require that all labor, supervisors, and third-party suppliers complete specific site-safety trainings. For infrequent site visitors, this may be limited to a half-hour or one-hour safety film pointing out the main potential hazards within the plant confines and identifying the planned escape routes. For personnel expected to be working full-time within the confines of the plant, there normally will be a four-hour, and sometimes eight-hour site-specific safety-training session. And more and more projects are also requiring all supervisors and craftsmen to have taken and passed the OSHA's 10-hour training course for the construction industry. The person preparing the job cost estimates must include all of these training hours as part of the job costs, even though they are not productive man-hours.

Finally, all major power plant construction projects include many heavy equipment lifts. Today, most owners and GCs require that the contractors who perform these lifts first prove, from an engineering perspective, that the lifts are possible and that they are safe. This can require concerted efforts by the lifting contractor to take measurements and prepare calculations. It can also require concerted efforts by the supervising contractor to review these measurements and calculations. This effort, on the part of both parties, involves people and their time. These costs should also be included in the job estimates. Even if the majority of this work is not done on-site, whoever does it burdens the recipient with a cost that needs to be captured somewhere, either in the overhead cost structure or as a direct cost to the job.

No matter the source of safety management efforts, they do incur costs. However, the small cost of these safety individuals is nothing compared with the cost of just one incident that could have been avoided had the proper safety measures been implemented. Safety starts at the beginning, with the person putting the estimate together. For more on safety, see chapter 9.

Other Considerations

When contemplating how to get the most work done for the lowest cost, the estimator will be taking on the role of both salesperson and site manager. As a salesperson, the estimator will try to come up with the lowest cost estimate possible in order to become the successful bidder. As a site manager, the estimator will be looking for the maximum flexibility to get the job done while staying within budget and schedule. The two roles are, in reality, at odds with each other. The site manager wants the maximum amount of money to do the work, while the salesperson wants the lowest price to sell the job. The estimator, therefore, is always in a quandary; how can he or she satisfy both roles?

One of the tools available to the estimator is to arrange for a brainstorming session with all of the project participants, as well as others who have been through similar projects. Out of a session like this can come a variety of cost-saving ideas. One of the first could be a review of component delivery. Should the components to be erected come in large, preassembled units, or should they come in knocked-down condition? Issues such as heavy-lifting equipment availability and the costs of this equipment could arise. Issues such as the trade-offs between using this equipment versus the time and cost of using on-site labor would be part of these discussions. Transportation costs from the shop to the site would be part of these discussions. And of course, the risk of damage in transit to large equipment components would have to be investigated. What does one do in the event of damage to an item like the one in figure 3–4?

Fig. 3–4. What happens now? (Courtesy of Construction Business Associates LLC)

Planning ahead how to handle large components may lead to fewer on-site man-hours, which will translate into fewer problems if there are labor shortages. It may also result in a smaller supervisory staff. Fewer workers require fewer supervisors. A smaller workforce requires a smaller staff of safety personnel. Larger components may require less welding on-site, which also leads to a smaller quality control staff. All of this staff reduction adds up. It can actually help both the sales effort of the bid and the construction effort of the site management team. As with many other specialized functions, rigging and lifting large, heavy components can be subcontracted to firms specializing in these activities. In addition to passing off the responsibility for preparing the rigging and lifting plans, which also require resources, the risk of something going wrong can be passed to the specialized subcontractor as well—a potential cost reduction and risk mitigation strategy, all in one.

Another efficiency method is to employ "lean construction" techniques. Originating in the Toyota manufacturing industry, lean construction, or as it more commonly referred to, "lean," is the next wave of construction methodology to pressure the industry to become more efficient. During TXU Energy's original plan to build 11 coal-fired power plants in Texas, its chairman and CEO at the time said that it would shave $35/kW off the cost of construction of these plants by implementing lean construction. At the time (mid-2006), this equated to 2.5% of a $10 billion undertaking. With numbers like these being tossed around by high-profile power industry executives, lean construction deserves a look.

Lean is a construction methodology that essentially looks at the construction process as a continuous flow of work activities, supported by material, labor, and other required resources availability. Instead of focusing on discrete WBSs, this process works by the participants working together as an integrated, supportive organization to focus on the total project plan. Although there is still a master schedule and a six-week look-ahead plan, it is the weekly work plan, "last planner" in lean terminology, that drives the process. The weekly work plan identifies what can realistically be accomplished in the coming week, based on labor and material availability, project status, and so forth. Arrangements are then made for that work to be done. The six-week plan is then updated accordingly.

All the project participants jointly agree upon the commitments for the weekly plan. The results are reviewed the following week, and adjustments made for the week thereafter. Progress is measured by the number of activities completed versus the number planned to be completed. It normally does not take long for project participants to stop committing beyond their capabilities when they have to face their counterparts the next week and explain why they were the ones who were the cause of not meeting the past week's goals. The goal is to make the planning process more reliable. Work is now not planned by the master schedule; it is planned by the project participants making detailed short-term plans based on material availability; realistic, proven labor productivity; and so on. It is a collaborative planning approach. As Greg Howell, one of the two cofounders of the Lean Construction Institute puts it:

> Lean construction produces rapid learning. Starting at the assignment level, understanding an assignment to my crew as a promise to the next crew drives a stake in the ground. Understanding where our preparation fell short at a detailed level makes it possible to prevent recurrence. Apply Plan-Do-Check-Act to the planning system itself. The investigation of causes can lead to improvements in logistics, operation design, and coordination across a variety of levels. In most cases, preassembly increases as workflow becomes more predictable, in part because there is no need to keep stores of smaller pieces close to the workforce to assure people can be productive when things go wrong. One unexpected consequence of this line is the significant reduction in contingencies hidden in every budget and schedule. More important is the reduction in accidents and injuries with many sites reporting a 50% reduction. We believe these results are due to the improvement in production system design rather than the motivation and training of the workforce.

Finally, Developing the Numbers

When one gets down to it, estimating is all about developing the numbers. Estimates or budgets based on estimates are used to determine if a project is viable, and they are all made up of numbers. Numbers are also the backbone of the estimates used by contractors to prepare the bids submitted for project opportunities. So it is of paramount importance that these numbers are the right numbers. These numbers must represent exactly what they are said to represent, without hidden extras. For example, the man-hours to accomplish a certain task must be exactly the man-hours that are expected to be used for that task, no more, no less. If any contingency is required due to a lack of clarity or due to unforeseen events, those contingencies should be accounted for as separate numbers. Let's say that a utility has set aside a fixed number of man-hours to open, inspect, and make repairs on a steam turbine. Then those hours should be the base estimate for that work. If this utility is unsure of the amount of work that may ultimately be required, then it could potentially add some historical percentage, say 20%, but that percentage should be carried in the estimate as a separate line item, identified as unanticipated work contingency. If the turbine rotor is to be shipped off-site for repairs, and if there is the possibility of bad weather delaying its return, these potential costs should also be carried as a separate line item and clearly identified. That having been said, let's delve into developing the numbers for the estimate or budget.

First, let's address the tools that are available to develop the numbers that will make up the estimate. They can be very simple and straightforward, or they can be very complex. However, the objective should be to use tools that fit the job at hand. The days of scribbling some numbers on the back of a paper napkin, making a firm handshake and getting on with the work are long gone. But that's not to say that a few handwritten calculations, backed up by references to historical data, cannot be adequate for some projects. But most likely, today's projects, especially the mega projects like the 1,000-MW power plant, will require much more sophistication.

Generally, complex spreadsheets are the backbone of many of the estimates prepared in today's business climate. Many organizations prefer to use their own, internally developed spreadsheet formats. This affords them the ability to see each project estimate from the same perspective. It ensures that during the preparation of each estimate, the estimator has considered the same conditions and applied the same tests for validity to each. It also allows them to roll up all the project estimates into a summary of potential projects in the bidding stage. These data are frequently used to forecast future workload and resource needs from a company-wide perspective.

However, there are many estimators who have their own spreadsheet formats. They may have developed them over a career spanning many years. This gives them the comfort that when they present their estimate, they can substantiate it and all of the conditions upon which it is based. This is not a bad thing. First and foremost, estimators must feel comfortable with the product of their labors. But when this is the case, frequently these numbers must then be converted to a standardized format that the corporate decision makers are accustomed to seeing.

But whichever method is used, it should be prepared with sufficient detail for a third party to understand the logic used when the estimate was prepared. It should also be structured to allow for comparing the actual job results with the estimated numbers. This is invaluable data that should be available for preparing estimates for future jobs. Although the actual job execution may not follow the logic established in the estimate, the estimating format should be structured such that certain categories can always be compared: total man-hours, labor costs, supervision costs, tools, equipment and consumables, and the like.

In today's world of doing more with less, and with the availability of high-tech number-crunching and data-manipulation software, there are now many third-party vendor programs available for preparing estimates as well. These programs, or tools, are often packaged as a series of modules. One module generally stands alone as an estimating tool. Follow-on modules may include feedback systems for collecting site data to be used for the next estimate. Others may exist for actual job-site monitoring and performing "what if" analysis. Still others sometimes allow for enterprise management. But for the purposes of the discussions here, the most commonly used packages are all intended to get more done, more accurately and quicker.

There are also various software packages available to enhance the drudgery of doing takeoffs. While these often work well for routine construction work, they are seldom as useful for preparing estimates for power plant construction work. Generally, the estimator ends up having to rely on bills of materials, drawings, and specifications to develop reliable takeoffs. For example, if an estimator is preparing the cost for preassembling waterwall panels of a large boiler, along with seal boxes, buckstays, and panel seam welding, there just is no suitable software that can do this reliably, over and over. The estimator normally has to walk through the work efforts, figure out the size of the crew for the job, and add up the man-hours required. The estimator would first envision a crew bringing the panels from a laydown area to some prefabricated work tables. Then, the estimator would see these workers precisely aligning and tack welding the panels, side to side, and maybe end to end as well. Next, the crews would be seam-welding the panel seams on both sides, and other crews would be welding the tube ends, if required. Then, there would be a crew or two placing and welding in position any seal boxes that might be required for sootblower or other openings. Finally, the estimator would see additional crews bringing buckstays from the laydown area, positioning them on the panels, and making the required attachment welds. (See figs. 3–3 and 3–5.) As is evident, this type of takeoff and estimating is not often amenable to the use of prepackaged software.

Activity	Craft	Hours per Shift ST	OT	Shifts	Man-hours
A. Preassemble Waterwall Panels					
1. Bring Panels to preassembly area					
	2 Boilermakers	8	0	4	64
	3 Trade Helper	8	0	4	96
2. Align Panels on tables					
	2 Boilermakers	8	0	2	32
	2 Trade Helper	8	0	2	32
3. Tack Panel seams					
	2 Boilermaker Welders	8	0	2	32
	2 Trade Helper	8	0	2	32
4. Tack tube ends					
	2 Boilermaker Welders	8	0	2	32
	0 Trade Helper	8	0	2	0
5. Weld Panel seams ("x" feet × "y" feet/hour)					
	4 Boilmakers Welders	8	0	5	160
	2 Trade Helper	8	0	5	80
6. Weld tube ends ("x" equiv welds × "y" equiv welds/hour)					
	4 Boilmakers Welders	8	0	6	192
	2 Trade Helper	8	0	6	96
7. Install Seal Boxes					
	2 Boilermaker Welders	8	0	2	32
	1 Insulator	8	0	1	8
	2 Trade Helper	8	0	2	32
8. Bring Buckstays to preassembly area					
	1 Boilermaker	8	0	1	8
	2 Trade Helper	8	0	1	16
9. Position and weld Buckstays ("x" buckstays × "y" hours/buckstay)					
	2 Boilermakers	8	0	4	64
	1 Insulator	8	0	4	32
	2 Trade Helper	8	0	4	64
TOTAL	40 hours/week	2 weeks duration			1,104
LABOR	14 Avg Craftsmen				

Fig. 3–5. Waterwall preassembly man-hours estimate

Regardless of what tools are used, they are only as good as the data that are provided. The WBSs, shown as items A.1, A.2, and so on, in figure 3–5, must reflect the work that is actually expected to be accomplished, and they must be tied together. If interface points are not addressed, the estimate will not be correct because scope will have been missed. The man-hours used to determine how long a task will take to complete must be based on some solid historical data. Or if there is no historical data, the empirical method used should be documented for future comparison to actual results. The costing data for converting man-hours into labor costs must be accurate and complete. Let's say that in the previous example, shown in figure 3–5, the estimator was now ready to apply labor costs to the estimated man-hours. These labor costs cannot be estimates; they must be accurate numbers. From the discussions earlier in this chapter, we know that if this is to be a union job, the labor costs are usually accurate because they are fixed by contract. If it is not a union job, then there will be somewhat of a guess involved, but it still must be very close to the rates ultimately paid when the work commences.

In this case, let's assume it will be a union labor project. To determine the costs to be applied to the individual worker categories, the estimator would have to obtain the latest contractually negotiated rates, which include more than just the wages. Specifically, the estimator would need to determine the base wage of the worker classification, then the agreed-upon fringes. In this case, fringes are items such as payments into the union welfare fund, pension fund, other retirement instruments, training and industry enhancement funds, and more. After determining all of these cost adders, which can easily add 60% or more to the base wage of a journeyman, there is another 20% or more of the wage for burdens such as FICA (Social Security and Medicare), federal and state unemployment taxes, workers' compensation, and corporate liability insurances (fig. 3–6).[2]

Typical Employer Labor Cost Components for Boilermakers (2012: Mid-West USA, Union)

Wages	Fringes	Burdens
$ 35.01 Journeyman	$ 7.07 Welfare Fund	6.20% FICA
$ 37.51 Assistant Foreman	$ 10.69 Pension Fund	1.45% Medicare
$ 38.01 Foreman	$ 3.14 Annuity	0.80% FUI
$ 30.01 General Foreman	$ 0.34 MOST and Common Arc	4.55% SUI
	$ 0.38 Apprentice Fund	6.45% Workers' Comp
	$ 0.30 Training & Educational Fund	3.00% General Liability

Fig. 3–6. Typical labor cost components (Courtesy of Construction Business Associates LLC)

In the current example, a boilermaker journeyman earning $35.01 per hour would cost the employer $69.71 once all of the fringes and burdens have been added in (see fig. 3–6), or almost double the base wage, per straight time hour worked, at least for the first few weeks. Some of the burdens stop after a preset amount of payroll, and if there is a very large payroll, this could be worth enough to take into account. But for purposes of this discussion, if we look at the waterwall preassembly example in figure 3–5, where there were 1,104 man-hours estimated for the work, the labor cost would equal $76,960. That's before any tools, equipment, consumables, supervision, overhead, and profit are added. Throw all that in as well, and the cost per man-hour will go way up, and the estimator must include it all.

If there is overtime anticipated, that needs to be plugged in. Usually these rates are anywhere from 50% to 100% higher than the base rate. However, all of the fringes and burdens may not apply. An accurate estimate requires that this be carefully reviewed and that the numbers reflect true anticipated costs. Then, as stated earlier, the small tools and consumables that the craftsmen use must be factored into the cost. For this example, about 15% of the labor cost would be a good number. That's an adder of $11,544. As for equipment, there will probably be a hydraulic crane to move the panels and buckstays from the laydown area to the preassembly area. So if this job is scheduled for two weeks, the crane will most likely be there for the same duration, even though it is not used all of the time. At a rate of about $2,000 per week, this would be an additional $4,000, excluding getting it to the site and returning it back to the rental vendor.

Other costs involve job-site office facilities, phone service, cell phones, print machines, tool room and staffing for the tool room, change rooms, sanitary facilities, welding machines, materials, and more. Although these costs can vary significantly from job to job, there is at least a 20% factor to be applied to the base wage rate. But because these costs can vary so significantly, they should be estimated separately instead of being treated as a percentage of the wages. As has been discussed previously, one component of these costs that is often overlooked is the cost of getting all of this to and from the site: the freight costs, which should also be a separate line item.

There may be other services such as preparing as-built drawings or preparing special outage reports if the project was a major overhaul or modification. This can sometimes require a few weeks of time by two or three individuals, temporarily living near the site. Their costs, including travel, hotel, per diem, and so on, should also be factored into the estimate. We'll assume they are not required here.

Then, there is the all-important supervision. As discussed earlier, there are many ways to obtain supervision, and the costs for each are different. But these costs do form a significant portion of the overall job costs, and therefore they must also be part of the job estimate. If overtime work is anticipated, some of these supervisors also will be required to work those overtime hours. The cost of these hours and any overtime differentials must be included. And then there is the administrative staff, especially on a large, long-term job. Costs for accountants, time clerks, runners, and a host of other support personnel need to be estimated and made part of the total budget. For the waterwall panel assembly job, these might add up to another 10% of the labor cost, or $7,000–8,000.

Finally, there is the corporate overhead reimbursement and profit that is expected by whoever does the work. There is no one formula that will cover this. Each contractor and each owner will have internal requirements for this. But let's assume a 10% overhead and 5% profit (all on the total job cost). Where does that put the contractor's sell price for this waterwall preassembly work?

Adding to the $76,960 direct labor costs the $11,544 cost estimate of the small tools and consumables, and including the $4,000 for the crane, an assumed $2,500 for freight on all tools, consumables, and heavy equipment, plus $7,500 for the supervision and administration allotted for this portion of the work, we have a total of $102,504. Factoring in 10% for corporate overhead (and this is often higher), we get a before-profit cost to the contractor of $112,754. Assuming a 5% profit margin requirement, the person preparing the estimate for this project would arrive at a "sell" cost of $107,629. (Note that some contractors do not calculate profit on overhead, since overhead does not contribute to the actual success of the project, but is just a cost of doing business.)

After completing the estimates for all of the WBSs of the project, the estimate should be summarized in a format that clearly shows the cost categories that the project will be measured against. This summary should include, as a minimum:

- Man-hours
- Labor cost
- Small tools and consumables
- Heavy equipment
- Materials
- Subcontractor costs
- Supervision days
- Supervision cost
- Safety

- Quality
- Contingencies
- Overhead
- Profit
- Escalation

Finally, the estimate should be accompanied by a schedule, a manpower loading chart, and a cash flow projection. It should also have, attached, the list of assumptions made as it was being prepared, along with a standard, detailed checklist of items to be considered during the preparation of each estimate, such as in Appendix A or B. Then, as a last item, it should be checked by a third party for arithmetical correctness and logic. The slip of a decimal or the incorrect copying of a formula in a spreadsheet can be disastrous.

T&M, Cost-Plus, and Change Orders

Most of the foregoing addressed estimates normally prepared for firm priced contracting where the contractor offers to perform the work for a single lump sum price based on the specifications and estimating assumptions. However, there are other forms of contracting. Two of the more common forms are T&M (time and materials) and cost-plus. And then there are times when changes occur that require re-estimates.

T&M

Time and material contracts are straightforward to develop. The estimator develops a set of unit rates for labor, tools, equipment, supervision, and so forth. These rates are then tabulated and made part of the contract, along with a stipulation that any material purchases would be passed on to the client at cost plus some specified markup. The difficulty is determining how to develop the rates and the material markups.

To be able to develop these rates and markups, the first question to ask is: What amount of work is expected, and what would its value be if it were firm-priced? To answer this, one needs to follow the process established above for developing an estimate. All of the procedures such as using proper estimating factors, making a site visit; crew sizing; labor sourcing; and anticipating the supervision, tools, equipment, and material needs must be followed. Although the firm price developed in this way does not have to be as exact as if it were going to be contractually enforced, it still should be very accurate. With this information now at hand, and categorized as discussed earlier, one can see how many man-hours are expected to be used, what the cost of these hours are, what the cost of the supervision hours are, and also the cost of tools, equipment, and the like.

Assuming the T&M contract is straightforward, there is a simple method for preparing the price for this type of structure. One adds up the indirect costs of the estimate, that is, the nonproductive costs such as overhead, profit, home office support, possibly some site services, plus safety and quality control, and spreads it across the direct costs. This would then increase the value of the labor cost, the supervision cost, the equipment costs, and so on, by the prorated portion of the indirect costs. Then, dividing the man-hours into the increased labor cost would result in new rates that reflect *all* costs anticipated to be associated with the labor component. The same thing would be done with the supervision, if not also included in the revised labor cost. For the tools, equipment, and materials, one would also now have the percentage markup needed for these items.

Cost-plus

To establish the "plus" for cost-plus contracts, one would follow the same methods as for T&M work, except the rates and other direct cost items would be the actual cost. The indirect costs would be established as a percent of the total direct costs anticipated for the work. This percent would then be the "plus." Let's go back to the example in figure 3–5, the waterwall preassembly work, where the 1,104 man-hours cost $76,960, and all of the other direct costs, such as tools, equipment, supervision, and so on, cost $25,544. This resulted in a total direct cost of $102,504. Then we assumed a 10% overhead and an additional 5% profit would be built into the sell price of that work. That is basically the cost plus 15%. But without a good understanding of the total job value, and hence the actual amount this 15% will bring back to the company, one may not be able to judge whether the adder is adequate.

Change orders

Most projects incur changes. Changes usually incur costs. Costs generally are recovered through a change order. But how is a change order estimated? If the job has pre-established T&M rates, or a pre-established cost-plus process to be used for changes, there may be nothing to do, unless the change significantly alters the work size or scope. But if a change order is required to be firm-priced, someone must estimate its value.

This can be done in several ways. One is to do a mini estimate, similar to the waterwall preassembly work discussed earlier. This would result in a change being treated as a mini job, with all of its attendant support and indirect costs. But another way to prepare a change order estimate is to add the work scope into the original project estimate as a separate WBS, and then rerun the base estimate calculations and note the bottom line change from the original estimate to the revised one. The decision on which is the better method is specific to each occasion. Since changes can be for increases as well as decreases, most contractors will go for the maximum when there are adders and offer the minimum give-back when the change is a reduction.

Summary

As was stated at the outset, estimating the construction work for a power plant project can be a daunting task. There is never only one right way to do this. The estimator needs to be all-knowing and think the job through in the manner the site management team will most likely get the work done. But the estimator also has to think like a salesperson to sell the job to the client or to the investors if the estimator is part of the owner's team. The estimator also has to understand the intangibles, those items that add cost yet do not add production, such as safety, quality, administration, and more. The estimator is expected to understand what will happen in the future, before the future ever gets here.

To do a good job of this, the estimator has to be field-worthy. The estimator has to be able to visit the job site and understand what will and what will not impact the job and, by extension, the estimate. The estimator has to understand where the craft labor will be coming from and understand how that labor performs. The estimator has to understand the cost of this labor and the trade-offs among OEM sources, third-party sources, specialized contracting, and direct hiring. The estimator also has to understand subcontracting and how to prequalify subcontractors that may be considered for the work.

A good estimator has to be in tune with the availability of skilled supervision. Similar to the unavailability of skilled labor, the unavailability of skilled supervision is also an issue, and the estimator must find sources and get those sources priced and committed to the project. The same goes for heavy equipment. Without this equipment being committed, the pricing used in the estimate may have very little relation to the costs eventually incurred.

Finally, preparing a realistic estimate requires realistic estimating factors, both for the development of the man-hours and for the costing applied to those man-hours. Additionally, the estimate should be accompanied by the schedule it is based on—actually, the schedule should be based on the estimate. It should also include all of the assumptions made when it was prepared and a completed checklist of standard items to be reviewed for any estimate.

References

1 Based on a 2012 study by the author. However, there are wide swings from state to state and contractor to contractor. For example, workers' compensation rates for one boilermaker contractor in New Mexico averaged $2.26 per man-hour, while rates for the same contractor's iron workers in Illinois were at $15.45.

2 Ibid.

More on the Planning Process 4

Chapter 1 was devoted to discussing the basics of how to go through the preplanning stage of a power plant construction project and move on to the actual planning. The discussions also included a road map of the various paths that could be taken to go from the preplanning process to the planning stage. A hypothetical 1,000-MW combined-cycle project was described for use as a standard model, and the discussions throughout chapter 1 were about this model.

However, the information in chapter 1 did not include many other topics that are also critical for ensuring a smooth and successful construction project. For example, there was no discussion on permitting, codes, or regulations. There was no discussion on personnel sourcing and training, for both the management team and the craft labor. Also, discussions on the sourcing of heavy equipment and small tools and consumables were not included. So this chapter will fill in those gaps.

Permitting

As any project manager who has been involved in any major infrastructure project clearly knows, the permitting process is daunting. It not only affects the ultimate timing of the project, it also affects the costs, the original parameters under which the project was conceived, and the ultimate success of the project. Success is defined here as completion on time, within budget, and with minimal scope changes, and permitting has an impact on all three.

Some permits take much longer than initially anticipated, sometimes stretching over a period of several years, often due to the enormous amount of testing and documentation required for submittal. Sometimes governmental legislation is changed, and the process has to be extended, or even begun anew. At other times, permit issuances require modifications to the original plant design, mechanically or operationally, adding to the cost and/or schedule. And then there are times when input from the public unexpectedly requires changes to plans stemming from traffic patterns, visibility aesthetics (think windmills—not in my backyard), and environmental issues.

What are some of the permits that are involved? Figure 4–1 shows a list of just some of the permits that may be required to retrofit an existing coal-fired power plant into a combined-cycle plant. As one can see, there are over 20 permits, and the list is not all-inclusive. Some of these are usually fairly straightforward, such as right-of-ways (if the land is public use), electrical, HVAC, plumbing, and health department. Others can be very time consuming, as noted above.

Some of the Permits Required to Proceed with the Site Works	
Air Permits · Air Pollution Control Construction Permit · CAAP (Title V) Permit · Acid Rain (Title IV) Permit · Open Burning Permit · Demolition Permits	Wetlands Impact Permit (US Army Corps of Engineers)
	Threatened and Endangered Species Permit (US Army Corps of Engineers)
	County Building Permit
	Health Department Permit
Wastewater Permits · NPDES Permit (clean water, etc.) · Water Pollution Control Construction Permit · Sanitary Holding System/Septic Permit	County Right-of-Way Permits
	Sate Right-of-Way Permits
	Electrical Permit
Determination of No Hazard to Air Navigation Permit (FAA)	HVAC & Plumbing Permits
	Demolition Permit
Pollutant Discharge Elimination System Permit	Noise Control Permits
	Erosion and Sediment Control Permit

Fig. 4–1. Required permits

Interestingly, some permits cannot be finalized until after the plant design, or redesign, is well along, which creates issues with projects using variations of the design-build project delivery system described in chapter 1. As an example, the requirements for obtaining some of the air permits are such that the information needed, such as exact particulate matter release, may not be known until the plant design is well underway. But if the site works are delayed until this information is available and reliable, the design-build delivery system cannot be used, possibly increasing the cost of the plant and negating its financial viability. Therefore, some of the permitting processes allow for phased applications. This does insert an element of risk into the project, a risk that the permits will not be issued unless costly rework is undertaken, but it will be a known risk that usually can be mitigated during the design stage.

Many of these permits require an initial application, with many, many pages of documentation. They may also require photographs, verified test results, and detailed descriptions of how compliance with the permit will be enacted and verified. They may also require input from the local populace, necessitating town hall meetings where the residents are afforded an opportunity to comment on the proposed work and permitting process. Sometimes this may require more than one town meeting, held several months apart. Therefore, when contemplating a large project, many owners hire a separate company just to manage the permitting process.

Codes, Standards, and Regulations[1]

As a project is being planned, the rules under which it will be developed, designed, manufactured, and built must be clearly understood. That's where the codes, standards, and regulations come in. The technical section of most power plant construction contracts will spell out exactly which industry rules are to be followed. In addition, there may be statutory rule requirements that apply even if the contract is silent on them, and there may be internal company policies that dictate additional compliance. This can be confusing, so it benefits all parties if the applicable codes, standards, and regulation requirements are spelled out in the contract, indicating whether they are contractual, statutory, or just company policy.

Not only do the specific code or codes need to be called out, but the specific sections of the code also need to be identified, and the applicable editions or case dates must also be specified. Usually, this is not an issue, but once in a while there are changes in the codes that may impact the work to be performed. In these instances, if the work has to be redone because the incorrect edition of the code was used, the contractor is usually liable for the cost of the rework as well as any additional costs to get back on schedule as a result of the rework.

So, what are some of these codes, standards, and regulations that may be encountered in a power plant construction project? The most common and complex usually will be the ASME Boiler and Pressure Vessel Code (hereinafter, ASME Code), and the National Board Inspection Code

(NBIC), both of which contain the rules with which vessels subjected to pressures above atmospheric are usually required to comply, at least in the United States and Canada. (ASME stands for American Society of Mechanical Engineers.) For example, a boiler in a power plant, whether it is fired or unfired, will be subject to these codes. Piping systems that carry pressurized medium will also be required to meet certain sections of the ASME B31 Code for Pressure Piping (e.g., B31.1 Power Piping, B31.3 Process Piping, B31.8 Building Services Piping, etc.). But how did this all come about? Well, let's read on.

ASME Boiler and Pressure Vessel Code and the National Board Inspection Code

The ASME Boiler and Pressure Vessel Code (ASME Code) and the National Board of Boiler and Pressure Vessel Inspectors' National Board Inspection Code (NBIC) are the codes that govern nearly every aspect of design, material, fabrication, assembly, installation, inspection, examination, testing, repair, and alteration of boilers and pressure vessels installed in the United States and Canada. In the broadest terms, the ASME Code governs new boiler and pressure vessel construction, and the NBIC governs the inspection, repair, and alterations of existing units. How did the two codes evolve, and what is their interrelationship?

In 1911, after almost 10,000 boiler explosions, ASME established a committee to formulate standard specifications for the construction of boilers and pressure vessels. In 1914, this committee established the first edition of what came to be known as the ASME Boiler and Pressure Vessel Code.

When first published, the ASME Code was divided in two parts. Part I covered new construction of power and heating boilers, and Part II covered applicable rules for existing installations. Today it is composed of 12 code sections, as follows:

I.   Rules for Construction of Power Boilers
II. Materials (Three Parts)
III. Rules for Construction of Nuclear Components (Three Divisions)
IV. Rules for Construction of Heating Boilers
V.   Recommended Rules for the Care and Operation of Heating Boilers
VI. Nondestructive Examination

VII. Recommended Guidelines for the Care of Power Boilers

VIII. Rules for Construction of Pressure Vessels (Three Divisions)

IX. Welding and Brazing Qualifications

X. Fiber-Reinforced Plastic Pressure Vessels

XI. Rules for Inservice Inspection of Nuclear Power Plant Components

XII. Rules for Construction and Continued Service of Transport Tanks

A new edition of each section is issued every two years along with periodic issuance of interpretations and code cases approved by the committee.

In 1919 the National Board of Boiler and Pressure Vessel Inspectors (the National Board) was formed. This board took on the lead role of promoting uniform boiler laws; provide a forum for the interchange of opinions; uniform stamping of boilers; and the uniform training, examinations, and certification of boiler inspectors. The 1946 Edition of the NBIC had six chapters. Today, it is composed of three volumes or parts, each with numerous subsections.

Part 1   Installation

Part 2   Inspection

Part 3   Repairs and Alterations

Like the ASME Code, new editions approved by the NBIC Committee are issued every two years with periodic interpretations issued between editions.

Today, every U.S. state and Canadian province has either a boiler law or pressure vessel law, and most have both. For an organization that designs, manufactures, or installs to certify work as meeting the ASME Code, the organization must first obtain a Certificate of Authorization to use the appropriate certification mark (formerly known as the code symbol stamp): S, A, PP, V, U, U2, U3, UV, N, NA, NPT, NV, and so forth. Likewise, to certify that a repair or an alteration to a pressure-retaining item meets the requirements of the NBIC, an organization must possess a Certificate of Authorization to use the R, NR, or VR stamp. However, the authorities of the jurisdiction where the boiler, pressure vessel, or pressure-retaining item is installed will have the final say on whether a boiler, pressure vessel, or pressure-retaining item is allowed to operate. Therefore, it is important to be clear on the rules and regulations in force in the locale where the pressure part–related work will take place, whether the contract language addresses them or not. In addition, many insurance carriers will also demand proof of certain code compliances before they will issue a certificate of insurance.

Other codes, standards, and regulations

Of course, there is much more to building a power plant than pressure vessels (e.g., boilers and piping). There is the concrete that supports most of the structure and the soil that supports the concrete. There is the steel that holds up many of the plant's components and equipment. There are all of the electrical wiring, cabling, and control devices that allow the plant to operate and communicate. And the list goes on and on. How does one know where to find the applicable codes, standards, and regulations that apply to all of this?

Some designers use lists that were developed in-house and provide them for inclusion in the construction contracts. Some general contractors do the same. Many lists are so long that they take up multiple pages in a contract, and no one will ever read all of them. Therefore, this usually becomes the responsibility of the quality management team. These team members ultimately are the ones who have to make sense of the codes, standards, and regulations that apply and then figure out how to apply them. More on how they accomplish this is in chapter 8.

For a glimpse into what these quality personnel are faced with throughout the project, here is a partial listing of the groups that promulgate some of these codes, standards, and regulations:

ACI   American Concrete Institute
AISC   American Institute of Steel Construction
ANSI   American National Standards Institute
API   American Petroleum Institute
ASCE   American Society of Civil Engineers
ASME   American Society of Mechanical Engineers
ASTM   American Society of Testing and Materials
AWS   American Welding Society

Note that these are just some of those starting with the letter "A." There are dozens more, and it is beyond the scope of this book to include them all.

Personnel Sourcing and Training

People. They are the key element of the project. Even with all of the technological advances during the past century of building power plants, whether the advances were in rigging equipment, welding methods, software for scheduling, or software for integration and communications, the plant still does not get built, repaired, or revamped without the people. Technology can take over some of the how-to functions, but the what-to-do functions are still in the domain of the people.

Whether building the power plant, retrofitting it, or performing a major overhaul, there are usually three pools of personnel that are involved. The first is the management. They often plan the job, they usually set it up, and they are responsible to manage it. Then, there is the labor, both skilled and unskilled, that actually does the job. They are the ones on the ground that make it happen. And finally, there are the third-party vendors, subcontractors, and original equipment manufacturers (OEMs), all of whom may be supplying equipment and materials or providing services.

With the constant focus on the ultimate cost of the project, it is important that the people selected to work on the project are efficient, knowledgeable, enthusiastic, reliable, and, most of all, loyal to the job. Many times one hears that if it were not for the people, there would be no problems, and the job would go smoothly. Well, if it were not for the people, the job would not get done. The people make the job, and they can break the job. So it is critical that the right people are chosen and that they are properly trained and properly managed.

But getting the best people, and getting them to perform at their best, is not always so easy. There are many obstacles. There are shortages of skilled people. There are shortages of supervisory people. Vendors often get stretched to the limit during peak outage seasons. Job-site pay scales are not always attractive. Available overtime may be limited. Weather may be a factor. And even if none of the foregoing applies, enticing the worker to do his or her best is often a challenge in itself.

The management team

The management team is the glue that holds the job together. Where do these people come from? How can one be sure they are the right people for the job? How can one be sure they'll understand the needs of the job and be able to get the labor to perform? These are often the questions that are in the back of the minds of those responsible to set up the job.

Between 1990 and 2010, there was a significant loss of construction management personnel with experience in the power plant construction business.[2] Many supervisors retired. Others left due to a slowdown in the industry. And unfortunately, not many new ones joined up. This resulted in a net "brain drain," a loss of tribal knowledge that is impossible to recover. Yes, a few of the old timers are still willing to come out of retirement and help with specific tasks such as complex rigging or to provide some welding support. But very few of them will sign up for a new, long-term construction project, or even a full 14-week plant overhaul. Thus, here is the problem confronting the powers that be: "Where do I get qualified supervisors to manage my site work?"

First, a description of the skills required must be prepared. Thought must be given to the technical requirements of the job: Will electrical work be performed? What about civil work and/or structural steel erection? Will rigging expertise be required? What kind of welding will be performed, will it need to be code certified, and will it be a major production or just a couple of high pressure piping welds? Is rotating equipment part of the scope, along with special alignment criteria? How about specialized knowledge requirements like turbine disassembly/reassembly or high-pressure piping cold-springing? Will there be a lot of instrumentation and controls work?

Once the technical skill requirements have been listed, the nontechnical skill requirements should be defined. Skills such as the ability to manage other supervisors or the ability to manage a multitude of craftsmen should be clearly spelled out. Any special skills that are necessary for the position must be identified, such as the use of certain field tools like a theodolite or the use of computers or handheld smart tools. If a second language is necessary or desirable, it too should be part of the job description. Also, the working conditions that will be encountered should be included.

Finally, the preferred background of the person who will fill the position should be highlighted. Maybe this person should have already attained some supervisory experience. Maybe there is a need for the candidate to have worked a specific number of years on a certain make of turbine. Usually there is a requirement for a specific level of educational attainment, although experience often can be substituted for some of the years of required education.

Once the parameters are spelled out, the actual job description should be formalized (see the first edition of this book for an example of a job description for a typical field supervisor who will be responsible for the erection of a boiler in a new power plant). There are several reasons for formalizing this. The first and foremost reason is to have a written document that can be used as a checklist for ensuring that all skills and requirements are covered. The second reason for a formal document is to provide the person performing the job with a clear understanding of his or her responsibilities. The third reason is to minimize confusion among others, up and down the management chain, as to who is responsible for what.

Having said all of the above, however, the job description should not be filled with minutia. The mark of good supervisors is knowing what needs to be done and thinking on their feet. Too many details will restrict their options for using their heads and place unnecessary burdens on the rest of the management team.

After settling on the job requirements and desired skill levels, the level and method of pay must be determined. The pay level becomes a trade-off between the amount budgeted for the job and the amount needed to attract the talent desired. Ideally, when the job was estimated, the estimators used realistic numbers for the salaries of the required supervisors, and if the market has not changed significantly, these values should still be adequate. However, if the time has stretched between when the estimate was made and the job starts, or if the market becomes tight due to other projects needing the same resources, there could be problems attracting the desired supervision while staying within the budget. For more on this, see the section on "Contingency Plans" in chapter 1.

Once the job has been described, the required skills for its performance identified, the pay scale determined, and the number of positions decided, it is time to look for candidates. Obviously, the first place to look is within the organization. But assuming there are no viable candidates there, a search on the outside must be initiated, often using word of mouth. The power plant construction industry is a tight-knit group. Often, suitable candidates can be located through contacting only two or three acquaintances. But this is not foolproof. It does not work every time, so backup plans must be developed. Here is where staffing companies can be of service.

There are many companies, all over the world, that offer "contingent staff" personnel for power plant construction work. When working with these organizations, it is almost imperative to provide a job description. They use this information to search their databases for candidates. Once they have identified several prospects, they usually offer to arrange an interview for final selection. However, if the staffing agency is already well known and has been used before, the agency can be asked to make the final selection and send the candidate directly to the job site. However, when this type of arrangement is used, there should be an agreement that in the event that the candidate does not work out, the staffing agency will find a replacement, at no cost to the employer, except for the travel expenses incurred.

What is involved if the supervisory personnel are directly hired and paid by the employer? Beyond just the usual payroll process, which may already be set up to pay the labor, there are a host of additional requirements that must be followed when hiring (and while employing) them. There are several U.S. federal and state laws that dictate these hiring and employment practices, as follows:

- Antidiscrimination. The antidiscrimination law basically states that an employer may not discriminate against any individual with respect to compensation, terms, conditions, or privileges of employment because of the individual's race, color, religion, gender, or national origin.

- Sexual harassment. This is defined as unwelcome sexual conduct, whether verbal or physical, when submission to or rejection of such conduct is used as a basis for employment decisions.

- Accommodation for religious beliefs. This regulation requires an employer to make reasonable accommodations to the employee's sincerely held religious, moral, or ethical beliefs.

- Pregnancy. Employers may not refuse to hire a woman on the basis of pregnancy, childbirth, or related medical conditions (unless the condition creates a hazard for herself or for her fellow employees).

- Equal pay. Employers may not discriminate between employees on the basis of gender by paying lower wages to employees of one sex than paid to employees of the other sex for jobs that require equal skills, effort, and responsibility, and which are performed under similar working conditions.

- Age discrimination. In the United States, law prohibits discrimination solely on the basis of age against any individual who is at least 40 years of age.

- Americans with Disabilities Act (ADA). The ADA prohibits discrimination against any qualified individual with a disability who meets the skill, experience, education, and other job-related requirements.

- Verification of eligibility. Employers must not hire persons not authorized to work in the United States—there are various ways to make this verification, both on paper and paperless.

- Retaliation is prohibited. Legislation prohibits retaliation against an employee who files a charge against an employer for a violation of any of the foregoing.

- Recordkeeping. Recordkeeping requirements are imposed on employers under a number of laws.

- Posting requirements. Employers are required to prominently post official Equal Employment Opportunity (EEO) notices to employees. It is recommended that both English and Spanish versions of these notices be posted.

Running afoul of any of the foregoing laws is not conducive to the smooth operation of the project. Most construction sites have very few secrets. When a violation occurs, word of it usually gets out. But to make matters worse, when rumors start that an action is pending by a grieved employee or candidate for employment, the truth often has a way of becoming exaggerated to the point of discontent among the workers on staff.

Finally, legislation in the United States provides for very steep penalties in the event of being found in contravention of most of these rules. Victims of intentional religious, gender, and disability discrimination can be awarded up to $300,000 per individual. For victims of racial or ethnic discrimination, there are no limits. Juries can penalize as they see fit! Since most power plant construction projects are already operating under limited budget constraints, incurring fines of these magnitudes can definitely ruin an otherwise successful project.

Finally, successful managers never stop recruiting the people they work with. They recognize that the supervisors they have hired are people, and all people have individual personalities. They all have needs and wants. Most like to be recognized for their contributions. Most like to be given the responsibility to manage their scope of the work and the latitude to decide how to do this; they do not like to be micro-managed.

It is important to realize that the supervisory personnel set the mood of the project; their enthusiasm, or lack thereof, is directly imparted to the craftsmen, which, in turn, reflects directly on the work they perform. If the supervision is unhappy or not supportive of management, the project will be an uphill battle.

The labor

If the management team is the glue that holds the job together, then the labor is the job itself. The same questions that must be asked about the management team must be asked about the labor: Where do these people come from? How can one be sure they are the right ones? How can one be sure they'll understand the needs of the job and get it done? And then, how does one get them to do it?

As with the construction management personnel, from 1990 to 2010 there was also a significant loss of labor skilled in the crafts required for the power plant construction business. Many workers retired. Others left due to a slowdown in the industry. Unfortunately, there were not many new workers. This resulted in a net skill drain, a loss that only new people and lots of training could replace. Generally, old timers are not willing to come out of retirement to work in the crafts. Not being so young anymore, they are usually less agile and more susceptible to pains and strains. They sometimes make good trainers or even supervisors, but for hands-on work using the tools, they are often not well suited.

When planning for the craft labor for a job, the process is somewhat different from sourcing the management team. Specific job descriptions are usually not prepared. If it is a union job, the union hall uses job titles to identify specific skill sets. For example, when the job site calls for 30 boiler tube welders, the hall knows that they must be certified for welding small-bore tubes, in accordance with ASME Section IX, and that they will be expected to be able to weld without undue reject rates—*undue* defined as "outside of the norm of the other welders." This does not have to be specifically written. Also, their years of experience seldom enter into the decision; often there are very young craftsmen working alongside of some with many years of experience. It is their ability to perform the task safely and efficiently that is important.

Although the process of selecting the craftsmen is not as complex as that of selecting supervisors, one aspect is more critical: their pay. Since labor wages often make up half or more of the total construction cost, even a few pennies saved per hour worked can result in significant savings on a major new power plant project. As with the budget for supervision dollars, when the job was estimated the estimators should have used realistic numbers for the costs of the craftsmen, including any union fringes or other uplifts. Differing from supervision, these numbers seldom experience major fluctuations over short time periods; if the workers are unionized, there will be collective bargaining agreements in place with specific numbers and conditions. Of course, if a lot of time has elapsed, those agreements may have been renegotiated, and then there could be significant changes.

However, sometimes labor shortages occur. Unfortunately, when this happens, there are fewer incentives to offer to the labor than there are for the supervision. One cannot offer permanent employment to the average construction worker. One cannot offer a guarantee of participation in a specific future project. Offering the workers the opportunity to bring their families, along with paying for the family's living expenses, is not usually an option, and it is the same with transportation. About the only two areas where some creative offers can be made are in the paycheck and in the work schedule.

In addition to offering overtime, which is sometimes limited by the budget and by the schedule, bonuses can be offered to entice workers to come to one job rather than another. Onetime sign up bonuses are sometimes effective, as are bonuses to workers who recruit buddies and bonuses payable after a predetermined period of satisfactory work. Job completion bonuses are popular, especially when they are tied to early completion.

Two relatively new approaches being used by some labor unions are "helmets-to-hardhats" recruitment of discharged servicemen and service-women and a teaming effort with the U.S. Department of the Interior's Bureau of Indian Affairs to recruit and teach Native Americans the skills necessary to work in the construction industry.

If the job is non-union, there are more places to look for people. A recommendation from existing workers is often an excellent resource. Often, the supervisory staff has a list of craftsmen they personally know and have worked with in the past. Then, there are the local municipal and state labor departments. They have a list of workers drawing unemployment, and they also have lists of workers looking for work even if they are not drawing unemployment. And if all else fails, there are the third-party labor brokers, companies that specialize in keeping track of available workers who have the skills required for many jobs.

Using labor brokers has advantages and disadvantages. Similar to using a staffing agency for supervisory personnel, using a labor broker for site-level workers can remove the burden of running a payroll. Brokers can do this on behalf of the contractor and bill only a fixed rate per hour worked. They can take care of all wages, payment procedures, benefits, and fringes, as well as all filings for payroll withholdings. On the other hand, the workers now have a first loyalty to the agency. They know that their livelihood is dependent on first satisfying them, and second satisfying the user. They know that once this job is over, it is the agency that will help them get the next job, not the contractor of the day.

If the labor is to be hired and paid directly, as with supervisory personnel, there are a host of additional requirements that must be followed. The following are the same practices that were spelled out earlier:

- Antidiscrimination.
- Sexual harassment.
- Accommodation for religious beliefs.
- Pregnancy. However, there are logical limits when involving field personnel.
- Equal pay.
- Age discrimination.
- Americans with Disabilities Act (ADA). As with pregnancy, there are practical limits when employing field personnel.
- Verification of eligibility. A word of caution is required here. Companies employing construction workers, in general, are often subject to inspection by U.S. Immigration and Customs Enforcement (ICE) for adherence to their regulations—be prepared.

- Retaliation is prohibited.

- Recordkeeping. Although this may be onerous when there are hundreds of workers, it must be done.

- Posting requirements. The EEO posters must be placed in conspicuous places like change shacks, lunchrooms, and other regular gathering places.

Again, governmental legislation provides for very steep fines in the event of violation of these rules. With upwards of 500 workers for some of the main contractors on large power plant projects, there are many more possible whistleblowers than supervisory staff, so extra caution is paramount. In fact, training the management staff, including foremen, in the ways of the above rules is strongly recommended. In the event there is an issue, being able to show that the site followed proactive procedures by training the staff will go a long way toward mitigating antidiscrimination claims, if any arise.

The training

In today's world of fewer and fewer experienced supervisors and craftsmen, training is becoming more and more important. Many of the larger contractors in this business now have ongoing training programs in which they train and update their permanent supervisors on a regular basis. Some require programmed training as a prerequisite for promotions. Some of the larger power plant owners do the same thing. Even the building trade unions are on board with this.

But when it comes to the craftsmen, training becomes a much more involved challenge. The aforementioned dearth of craftsmen, trained or not, has had a big impact. That 20-year period when there were few entrants into the construction workforce, union or otherwise, was a period when training was minimal—there were few to train. Now there are various efforts underway to bring more candidates into the construction industry.

There are efforts such as corporations working with high schools and trade schools to introduce students to the construction industry. There are similar efforts at the two- and four-year college levels. There are companies that set up their own training programs, sometimes for specific projects and sometimes for future needs. Organized labor has been at the forefront of a lot of training as well.

With the advent of the Internet, many opportunities are available for persons interested in learning the trades to build or rebuild power plants. However, none of this substitutes for the hands-on learning that ultimately is needed to get the work done. This is where owners, contractors, and labor supply organizations sometimes work together to provide a hands-on training component. And when this is not feasible, each group must do what it can to train its own personnel for its own needs.

One such example is the training program in place in the UA (United Association of Journeymen and Apprentices of the Plumbing and Pipe Fitting Industry of the United States and Canada, more commonly called the Pipefitters Union). Although not the only labor organization to provide extensive training to its members, the UA's program has been around, in some fashion, for over a century. Their program takes an apprentice from day one through a five-year period that is made up of classroom and hands-on training. The classroom segments cover basics such as mathematics, mechanical drafting, and related science. The hands-on training is a combination of on-the-job training and specialized skills instruction at one of the Pipefitters Union halls, where there are sometimes facilities designed specifically for training. Figure 4–2 shows a mock-up used by the UA for teaching specialized rigging skills.

Fig. 4–2. Rigging training (Courtesy of the UA)

However, rigging is just one of the specialized skills needed to build power plants. Another, welding, may be even more critical. This is a skill that cannot be learned in the classroom. Although the metallurgy behind joining two parts together through the welding process is important to understand, it is the technique of how to actually do this, hands-on, that is critical. In the early days of power plant construction, this was mostly taught through on-the-job training. However, as power plant owners have become more cost conscious, the building trades have started their own welder training programs, away from the job site. The UA is very involved in this, as is the International Brotherhood of Boilermakers. Both organizations' programs are in compliance with the requirements of the ASME codes described earlier in this chapter. Figure 4–3 explains the cost savings to the industry from the boilermaker craftsmen off-site certification program, Common Arc. Combined with similar efforts by the UA and others, the efforts by the industry to have trained craftsmen readily available really do impact the bottom line. This is preplanning at its best!

And finally, for those skeptics that are unsure of the intrinsic value of training, studies by the Construction Industry Institute (CII) in 2007 have shown that craft training can generate big cost impacts. They found that investing only 1% of a project's labor budget in training could have double-digit returns in productivity, reduced absenteeism, and need for rework.[3]

Third-Party Vendors, Specialty Subs, and OEMs

Even with all of the supervision and with all of the labor, trained or not trained, there are some services that an owner or contractor will prefer to outsource. There are many reasons for this. Some services require unique skills that are only performed once every five or more years, so it is not practical for individual owners or contractors to train their personnel to perform these functions. Sometimes, there are risks associated with a particular activity such that performing the activity would impact the insurability of the contractor; subcontracting it to a specialty vendor puts the risk with the company best suited for managing it. Some services require personnel with access to proprietary or OEM knowledge that is often unavailable to personnel other than the OEM's staff. And then there are times when it is just simply more cost-effective to outsource the work than it is to perform certain construction-related activities with in-house resources.

Common Arc — A Power Plant Industry Cost Reduction Program

Until the late 1980s, welders hired by a contractor working on a boilermaker job were required to take an individual welding proficiency test, for each contractor for whom they worked, to ensure that they were qualified to weld for that specific contractor. Every time the welder moved to a different contractor, he was required to test again, for the new contractor. Sometimes, they even re-tested for the same contractor. Calculations by the industry at that time put these testing costs at around $500 per weld test.

In those days, as is still common today, when a large outage started, it was not uncommon to need 100 ASME Code-certified welders at the peak of the work. With test passing often hovering in the 50% range, this meant 200 welders would need to be tested before 100 were certified to work. In the 1980s, that added up to an enormous $100,000 per job, all of which took up time and was non-productive. But even more so, when multiplied by two outage seasons per year and by the number of plants that were having regular outages, it was calculated that the industry was spending over $55 million each year for all of this testing! So the boilermaker union and its contractors got together to see if they could devise a way to reduce these costs for their clients—the plant owners of the day.

In 1988, they established the Common Arc Corporation, a not-for profit organization chartered to reduce the cost of providing certified welders for member contractors and owners. Working with the ASME Code committee, Common Arc devised a simultaneous testing system whereby many welders could be tested and witnessed by many contractors, all at the same time. They designed this to be done at off-site locations, during slow work periods, so as not to interrupt the outage seasons, and they called it "Simultaneous Testing."

Through the use of this new program, today's cost of testing and having available a certified union welder has come from what would cost $1,000 to under $35 per man. Total industry costs now run under $2 million per year and there is a pool of approximately 11,000 certified welders that are available, instantaneously and without additional expenditures, for any outage work a Common Arc member may require. That's a cost saving measure that can only come from forward-thinking team work.

Fig. 4–3. Common Arc—a power plant industry cost-reduction program

Third-party vendors

Third-party vendors are often thought of as outside shops that provide specialized services, tools, or supplies. A very common example is a vendor providing services for turbine rotor repairs or generator rewinds. In the past, work on major equipment such as this would be performed in the vendor's shop. The equipment would be moved out from the plant, loaded on trucks and/or railroad cars, shipped to the vendor for repair, and then brought back and reinstalled in the plant. Since it was impractical for the plant, or almost any contractor, to have the equipment and personnel available to do these major repairs on-site, there really was no other choice than to hire a third-party vendor.

However, this did create an additional risk for the project—a risk of something going wrong and then the machine not being brought back per the schedule. A problem might occur at the vendor's shop, where there could be a breakdown in equipment, a labor action, or an overload of work. The problem might be encountered during transportation to or from the shop. Trucks break down, bridges wash out, and roads become impassable due to snow or rock slides. The problem might be obtaining and coordinating transport permits. Or the problem could be in the plant itself, either with the removal of the machine, or with its reinstallation.

With the increasing focus on cost and schedule, more and more outage supervisors now want this kind of work to be closer to their control. They prefer for this work to be done within their facilities, preferably in place so that many of the potential risks just described can be eliminated. This then requires the vendors to have specialized equipment for doing this kind of work at their clients' facilities, and it requires them to train their personnel in the use of it. The cost for the service is often higher than if it were done in the shop, but the risks to the schedule are reduced.

Specialty subs

There are many instances where it is more cost-effective to subcontract a complete part of the plant construction or repair. The rationale may be the unavailability of qualified personnel. It may be that the subcontractor can perform the work for a lower cost than the contractor doing it. It may also be that the risks inherent in the specific work scope are not acceptable to the contractor's insurer.

An example of this last case is the installation, maintenance, and removal of scaffolding. Most scaffold work is for temporary access for personnel to get to areas not normally accessed. Because of the temporary nature of scaffolding, most insurers rate the potential for accidents involving personnel on scaffolding higher than they do for much of the other work. When an accident does occur, the cause or the blame is often associated with the erection or maintenance of the scaffolding itself, which makes the installer of the scaffolding liable. Since there are contractors who specialize in the erection and maintenance of scaffolding, and who are certified and insured specifically for these types of risks, many owners and general contractors subcontract this specialty to them.

A similar situation exists with the removal or abatement of hazardous materials such as asbestos insulation. Due to insurance requirements that require specialized training for workers handling asbestos, most contractors will not work with asbestos. They use specialty subcontractors, thereby shifting the risk to those who have the training and experience to shoulder these risks. It is basically a matter of economics; the specialty contractor can usually perform the work scope more cost-effectively than a general contractor can.

Original equipment manufacturers (OEMs)

In contrast to third-party vendors and specialty subs, OEMs offer a type of support very few of the other parties can provide: the original drawings, specifications, calculations, and even the in-house engineering staff that is specific for the part, machine, or equipment being installed or repaired. Short of reverse engineering a replacement part, especially if the equipment it is in is still fairly new, going to an OEM is sometimes the only way to acquire it. Even when considering reverse engineering, there may be an issue with the time required to engineer, manufacture, and ship the part. There may also be an issue with warranties, because the OEM will certainly not warrant reverse engineering.

OEMs also offer technical personnel who are trained specifically on their product and equipment. These technical people have access to the engineering drawings and to engineers who can answer their questions and resolve problems when they arise. OEMs usually know the best practices to be used when erecting, disassembling, and reassembling equipment they have manufactured. They can often offer cost-saving ideas during these operations, and they almost always will guarantee the performance afterwards.

However, they cannot offer to warrant the workmanship of another contractor's labor. Therefore, engaging the OEM's contracting arm is a frequently used method for maintaining performance warranties for the installation of a new piece of equipment, whether it is a turbine, a boiler, or a piece of pollution control equipment. The OEM may not actually use its own in-house contracting division, but may subcontract the labor to a third party. From the owner's and general contractor's perspective, the risk is still with the OEM, and the owner and contractor are protected. Again, it is a matter of dollars and cents. Due to overhead, the OEM is frequently more expensive than a construction-only contractor, but it is often worth this cost differential to be able to shift the risk of incorrect or poor installation.

Equipment, Small Tools, and Consumables

Equipment

Deciding how and from where to obtain large pieces of construction equipment requires a multidisciplinary effort. Large construction equipment, such as cranes, bulldozers, heavy haul trucks, and so forth, is expensive. Usually the constructing entity—whether the owner, general contractor, or subcontractor—does not have this type of equipment. This type of equipment is generally the purvey of a specialty company. Arrangements can be complex for this equipment to be available when called for in the construction schedule. There may be other projects that already have contracted for it, or there may be issues of adequate sizing, such as needing an extra boom or longer jibs for a crane. It is because of the uniqueness of heavy construction equipment that other avenues of approaching the task to be done should be explored.

Maybe, instead of using heavy haul trailers to move large components within the project perimeter, such as bringing a preassembled heat recovery steam generator (HRSG) into position, consideration should be given to laying a railroad spur to the location. Maybe, instead of using two large cranes to raise a heavy boiler steam drum, consideration should be given to using special jacks that "pull" the drum up by the cables attached to it (fig. 4–4). The economics, including impact on schedule, must be closely reviewed along with a risk-analysis and backup plan in the event things go awry.

Fig. 4–4. Drum raising with hydraulic jacks (Courtesy Barnhart Crane & Rigging Company)

Small tools and consumables

When a tool or a consumable such as a special welding rod is not available, the craftsman may have to stop working. As part of the planning process, it becomes crucial to know where to find the small tools and consumables that the craftsmen will use, how to get them to the site in a timely manner, and how to control them once they are there.

Imagine a scenario where the superheater elements are being replaced in a large utility boiler. Envision them all hung out, tacked in place, and ready for final welding. Then assume that there are five pairs of welders, on each 10-hour shift, with 1,400 stainless steel welds to complete. This is a two-week, two-shift welding operation that will require approximately half a ton of welding rod. But suppose that halfway through the job, the welding supervisor finds out the last 500 pounds of rod is not stainless, it is plain 6010 carbon steel. The work has to stop, someone has to scramble to locate more rod, and two 10-person welding crews are standing around, getting paid to do nothing. That's expensive. But what's worse is that when the missing rod does finally arrive, more overtime may be required to avoid completion penalties, or in spite of the extra overtime (OT), there may still be penalties—maybe up to $500,000 per day! The point here is to realize the impact that the lack of planning can have even for mundane items such as small tools and consumables.

Summary

In chapter 1, the reader was taken from the initial decision-making process to deciding if it made sense to move forward with the project at hand. This was the start of the preplanning process. Then, the reader was taken through a lessons-learned process and into a matrix of decision making to determine the most apropos project delivery structure. From there, budgeting was discussed and contingency plans were addressed. With all of that and with the information in the subsequent chapters on bidding and putting the numbers together, the average power plant construction professional could be well on the way toward getting any project off the ground.

However, that information, in and of itself, is not all that is required to launch a project. Missing are a host of other issues such as permitting, codes, rules and regulations, personnel sourcing and training, subcontracting, and planning for the tools and equipment necessary to accomplish the work. These were addressed in this chapter.

The chapter started with an overview of the magnitude of permits and the time involved in the permitting process. Issues such as technical and operational requirements were addressed. Also addressed were third-party interests, such as unsightly wind turbines, which can delay or derail the permitting process.

A stroll through the world of codes, standards, and regulations followed, with a bit of background on how the codes affecting power plant pressure-containing vessels came about—after almost 10,000 boiler explosions in the late 1890s and early 1900s. The resulting ASME Boiler and Pressure Vessel Code, known as the ASME Code; other ASME codes; and their cousin, the NBIC, are now in use by every state in the United States and all provinces in Canada.

Since no discussion on planning would be complete without addressing the sourcing of resources, the remainder of this chapter went into personnel sourcing and training; a bit on subcontracting; and some ideas to be considered before sourcing equipment, tooling, and consumables.

Just like scope determination, equipment selection, scheduling, and budget preparation, personnel selection has to be a planned and well-executed process. A good construction job requires good construction people. Care in selecting the supervisory staff is of utmost importance. They set the tone of the project. Selecting the labor is also crucial, and there is often less flexibility in sourcing labor than supervision. The treatment of hiring both the supervision and the labor is governed by some very specific guidelines that must be followed to avoid serious legal and financial penalties. The project management personnel and the site supervisory staff must all be familiar with these guidelines. In the interest of space, the management of the supervision and labor was not addressed. For more on that topic, please see the first edition of this book.

Rounding out the site workforce are the third-party vendors, specialty subs, and OEMs. These organizations are hired for the specialized expertise they bring, expertise that most owners and general contractors do not have. They fill unique niches in the power plant construction industry because of their specialized services, such as turbine reblading or rotor rewinding. They may specialize in services that are too risky for a larger contractor to undertake, such as scaffolding or asbestos removal. They may be the OEM of certain equipment, and therefore they may be the only group with drawings, calculations, and engineering expertise to offer a guarantee of performance.

Finally, no one could get anything done without equipment, tools, and consumables. But the efficient sourcing of these must also be well planned. Large construction equipment can be expensive, so it is important to look at alternative ways to accomplish certain tasks to reduce this cost.

The same goes for the smaller tools and the consumables. Their sourcing needs to be planned in advance. An example was shown in which a team of welders had to stand around when they ran out of a specialized type of welding rod. A failure to properly plan can therefore become very costly. And if the contractor is subject to liquidated damages, the costs can be even greater.

In chapters 7 and 11, all of this previous planning will be pulled together. In those chapters it will become obvious that without this preplanning and planning, effective job-site management would be nearly impossible.

References

1 Supplementary comments included from personal communications on February 5, 2013, with J. T. Pillow, member ASME Standard Committee—BPV I Power Boilers, member Committee on National Board Inspection Code, co-author of the second edition of *Power Boilers: A Guide to Section I of the ASME Boiler and Pressure Vessel Code*, and contributor to the 4th edition of the *Companion Guide to the ASME Boiler and Pressure Vessel Code, Volume 1*.

2 Shuster, Erik. *Tracking New Coal-Fired Power Plants*. National Energy Technology Laboratory. February 18, 2008. http://www.energyjustice.net/files/coal/netl/2008-2-18.pdf.

3 Rubin, Debra K. "CII Study Shows Craft Training Can Generate Big Cost Impact," *Engineering News-Record*, August 20, 2007.

Terms and Conditions 5

erms and conditions (T&Cs) are the legal details of liability and responsibility that govern the contract. They are the statements, clauses, paragraphs—sometimes pages and pages—that one hopes will never be needed. But without them, chaos may result, especially in today's litigious climate. Usually, they are a separate section of the contractor's proposal and/or contract. They are mostly written in legalese and therefore too frequently subject to multiple interpretations.

This chapter will show why T&Cs are needed. It will discuss the legal framework that usually surrounds them. It will delve into which ones are must-haves in a power plant construction contract and which are just nice-to-haves. Some of the must-haves that will be discussed are consequential damages, limits of liability, warranties, changes, delays, and dispute resolution. Some of the nice-to-haves that will be discussed are default, suspension and termination, time-is-of-the-essence, extra work, and escalation. Examples will be provided to emphasize the importance of including or excluding the specific term and condition. Some of these examples will show the perspective of the purchaser/owner, while others will view it from the supplier's or contractor's point of view. A table of comparison among all of the parties is included to provide an understanding of each other's position.

Finally, a word of caution is in order. Contract legalese requires careful preparation. It is not for the faint of heart. If there ever arises an occasion where the T&Cs of a contract are required to resolve an issue, they have to be very clear. They have to be enforceable in a court of law. And they have to provide all of the protections the company requires, within the confines of the laws in the jurisdiction where they are being used. This means that expert help should always be used in their preparation. Trying to go it alone when preparing a contract is a risky effort fraught with the potential for disastrous consequences.

The Legal Framework

For a contract to be an effective document, it must be enforceable in a court of law. Therefore, when a contract is drawn up, it is drawn up under the auspices of a legal jurisdiction. In the United States, this is usually in accordance with the laws of a particular state. For example, there may be a clause that states that any disputes will be dealt with in accordance with the laws of the State of New York. This has no bearing on the laws under which the companies are incorporated or the jurisdictional laws where the work is to be performed. It is simply an agreement between the parties to the contract that in the event they cannot resolve differences in the future, they will rely on the rules and regulations in effect in a specific jurisdiction, such as a state.

Although most contracts will specify a legal jurisdiction, there may also be additional agreements that either supersede certain specific rules or add to them. For example, if there is an arbitration clause, it may specify that arbitration be in accordance with the rules and regulations of the American Arbitration Association, regardless of what the contractual jurisdictional rules require; that is, those rules may be superseded.

The Terms and Conditions of the Contract

The legal rules that govern the liabilities and responsibilities of the contracting parties are collectively called the terms and conditions (T&Cs) of the contract. These rules are put in place by all of the parties to the contract to protect themselves in the event of disputes. During the contract negotiations, each party jockeys for position, looking to secure the most favorable terms they can for themselves, especially if issues of dispute end up being decided in a court of law.

Must-have

Although no one contract format is adequate for all projects, a typical contract's T&Cs should address, at a minimum, the following (the must-haves), not in any order of preference:

- Consequential damages
- Limits of liability
- Warranties—expressed and implied

- Remedies available "at law or in equity"
- Third-party claims
- Changes
- Delays
- Applicable law
- Dispute resolution
- Indemnification
- Proprietary and commercial property protection
- Performance guarantees

Let's explore the meaning and use of each.

Consequential damages. Damages that are not quantifiable, and therefore normally not insurable, are considered consequential. They are costs one step removed from direct costs. They may include lost profits, loss of bonding capacity, loss of business reputation, and more.

For example, a contractor's crane boom falls and takes down the power lines between the main transformer and the switchyard. The direct damage to the client is the damage sustained by the power line. The consequential damage could be the loss of profits or revenues due to business interruption since the plant will not be able to dispatch power from the unit until the power line is replaced.

Standard contracting practices today include a reciprocal waiver of consequential damages between the owner and the contractor. The waiver should be written in a positive vein, stating what is included. For example, a waiver may state, "The Owner hereby waives any form of incidental or consequential damages including but not limited to the Owner's loss of use of the facilities, loss of income, profit or financing related to the Work, cost of replacement power, loss of tax credits, or any other indirect loss arising from the conduct of the parties to this Contract; and the Contractor hereby waives any form of incidental or consequential damages including but not limited to damages incurred by the Contractor for loss of financing, business, and reputation, and for loss of profits except anticipated profits arising directly from the Work, or any other indirect loss arising from the conduct of the parties to this contract."

According to the U.S. Uniform Commercial Code, if consequential damages are not specifically excluded from a contract, in writing, the contractor can be held liable for these damages. Mutually excluding consequential damages is one method of mitigating this risk. By waiving claims for consequential damages, the parties limit their exposure to direct damages.

So what does this mean in the example with the crane? It means that by waiving the right to claim for consequential damages, the owner has removed a potential nonquantifiable risk from the responsibility of the contractor. This allows the contractor to not include contingency monies for potential consequential damages when bidding for performing the work, which ultimately results in a lower price to the owner for the work.

Limits of liability. Prudent contracting includes placing limits on the total exposure to the overall liability that can be incurred in a contract. In addition to limiting a contractor's exposure to some portion of the contract value due to schedule overruns (if, in fact, the contract imposes a penalty for overruns), most contractors go beyond just protecting themselves from delay penalties to also limiting their total exposure from any and all occurrences to an aggregate dollar amount. This is usually highlighted by a separate limitation of liability clause that essentially states that the contractor's overall liability from all causes is limited to some specified value, or some specific percentage of the contract value.

A limitation of liability clause is a contractual provision that restricts the amount of damages a client can recover from a contractor. Properly drafted, and where enforceable, it can provide protection against contractual breaches and negligence and is a valuable tool for allocating project risk. But it does not protect the contractor against liability for intentional misconduct, nor does it limit the contractor's liability to persons other than the client. Third parties who have not signed the contractor's contract are not bound by the limitation provision.

The limitation of liability clause must be carefully drafted to cover the types of risks that could be encountered and the legal theories the claimant could assert. Otherwise, a court may determine that the clause, although valid, does not apply to all or some of the action. Therefore, a contractor would want to be sure to include wording that covers potential damage to the client's property and liquidated damages as well, if they exist.

A confusing issue in clauses that limit liability can be the cost of litigating. With wording similar to the following, this issue is made somewhat clearer: "In any action to enforce or interpret the terms of this Agreement, the prevailing party shall be awarded, in addition to any other remedy or compensation, its attorneys' fees and costs, including fees of expert witnesses."

Although most jurisdictions permit limitation of liability, some do not. To avoid invalidation of an entire agreement or an entire limitation of liability clause, a savings clause, such as "To the fullest extent permitted by law," should be used where multijurisdictional use is anticipated. Better yet, review the limitation of liability clause with a good construction lawyer in the jurisdiction where the clause will be used.

To give this clause extra teeth, many contractors also make it a supremacy clause by adding the following, all in capital letters: "THIS CLAUSE GOVERNS OVER ANY CONFLICTING CLAUSES FOUND ELSEWHERE IN THIS CONTRACT." The intent is to keep a contractor from going out of business over just one failed contract.

Limitation of liability clauses can be effective tools to redress the economic imbalances present in large power plant construction projects. However, they must be carefully and clearly drafted. Further, the contractor's contracting and management practices must be sensitive to the limitation clause. Careless practices, such as allowing contract assignment to other entities, may undermine the clause and waste the effort incurred in negotiating the limitation clause.

Warranties—expressed and implied. Warranties can be directly expressed in a contract, or they can be implied. Expressed warranties are exactly that: They are expressed, or stated directly, in the contract document. For example, "The contractor *warrants* that there will be no boiler tube leaks for 12 months after completion of the work, provided the unit is operated in accordance with original equipment manufacturer (OEM) recommendations." Or, "The contractor will return the area to its preexisting condition upon completion of the work." A nonexpressed (or implied) warranty, on the other hand, is a warranty not intended by the contractor. For example, there may be remedies available to the parties to the contract based on legal precedence. Unless specifically addressed in the contract, they are implied, and the prudent contractor will include a statement specifically excluding them.

It is important for all parties to the contract to understand the exact promises, or warranties that are being offered, and under what conditions they are valid. It is also important for all parties to understand what is excluded. That is why most contracts will have rather lengthy warranty articles. They will have details about what is expressly warranted, such as the no-leakage phrase above, along with any conditions that may be applicable. In this example, if the boiler were operated under conditions not foreseen or designed for, such as daily up and down cycling when the design was for base load, the owner/operator may not be able to force the contractor to return and repair a tube leak.

The implied warranty issue is more complex. First, it usually arises over something that is not spelled out. Let's look at a couple of examples:

Example 1. A civil works contractor was hired to pour the slab for the permanent plant warehouse. Upon inspection, the building inspector decided it was not up to code. The contractor was asked to cut out some sections and add footings beneath the slab for some columns that would be installed next. This requirement was not on the plans the owner had given to the contractor. The contractor, of course, expected to be reimbursed for this extra work. The owner, on the other hand, said it was warranty work, and that the contractor had violated his implied warranty.

How does one resolve such an issue? The first question to ask is did the contractor know, or could the contractor have known, that there would be columns that required footings? If the answer is yes, then the contractor should have questioned the owner about the need for footings. If the answer is no, and especially if the owner used an architect or engineer to design the slab, then the owner generally would be required to reimburse the contractor for this extra work.

However, to minimize this kind of issue in the first place, wording similar to the following is recommended: "The contractor is not responsible for work that does not comply with the building code if that work complies with the building plans and design *that were provided to the contractor*." In other words, contractors need to limit their implied warranties by stating, very specifically, that they are relying on the owner's plans. This adds teeth to the legal principle that plans and specifications provided by an owner contain an "implied warranty" of accuracy.

Example 2. The owner provides a supplier with the requirements for a baghouse. The supplier provides the contractor with the specifications for erecting the baghouse on-site. The owner-supplied requirements were performance-based, specifying the results to be obtained and leaving it up to the supplier to design for attainment of these results. The specifications provided by the supplier, however, were design specifications, explicitly stating dimensions, locations, sizes, and types of materials. Deviations were not permitted without supplier approval.

The requirements that the owner provided to the supplier were essentially a performance type of specification. This type of specification carries an *implied* assumption of risk by the supplier that it will select the proper equipment and materials to accomplish the task. Therefore, if the equipment does not meet the performance criteria spelled out in the owner's performance specification, the supplier will be held liable to correct this, not the owner.

On the other hand, the specifications provided by the supplier to the contractor were design specifications, which carry an *implied* warranty that if they are followed, an acceptable result will be produced. Therefore, if the contractor builds the baghouse in accordance with the specifications, and if it does not function as the supplier intended, the contractor will not be liable, the supplier will be.

Although these two examples are straightforward, life is not usually so simple. The distinction between design and performance specifications is not always clear. Some construction contracts have elements of both. Once issues like these have occurred, identifying the relevant factors requires a close review of the contract and the contractor's particular difficulties. Therefore, it is best to plan ahead when preparing specifications and contracts, and adopt wording that is unambiguous.

There is one caution. If the work is being performed under a design-build scenario, the implied warranty concept discussed above loses its effectiveness. In contrast to design-bid-build contracts, design-build contracts require that the contractor be responsible for preparing the plans and specifications as part of the overall contract performance. Therefore, the risk and accuracy of the plans and specifications are transferred to the contractor, and the owner's implied warranty of constructability is gone or substantially diluted.

In contracting, a typical tactic by owners and upper-tier contractors is to use a disclaimer in an attempt to shift the risk of adequacy of the plans and specifications. One way is to state that the specifications are advisory only, or solely for the convenience of the contractor. Another common phrase is that the owner does not assume "any responsibility for the data as being representative of the conditions and materials which may be encountered." There have been cases where this has been upheld, disallowing recovery for additional costs incurred by a contractor who encountered unforeseen difficulties due to soil conditions that were not anticipated.

And then, there is the reality of the facts in a case. When an owner's specification includes a statement requiring the contractor to examine the site, check the plans and dimensions, and assume responsibility for conflicts, the contractor has an obligation to look, see, and measure. A case in point involved a rolling door for a warehouse. The door was shown to be 16 feet wide. But when the contractor went to install it, it would not fit because the opening for the door had an obstructing column. The contractor claimed that the owner-provided specification was defective— that the owner had breached the implied warranty of sufficiency of the specifications. However, the courts held that the implied warranty was negated by the look, see, and measurement clause, which, if prudently followed, would have shown the conflict.

But regardless of previous judicial rulings, there is no guarantee that the next time a similar situation arises, the ruling will again be the same. Courts in different states do not always see eye-to-eye on issues. Judges do not always agree, and if an issue goes to a trial by jury, the outcome is never guaranteed. The United States federal court system is also not obligated to follow previous state court rulings. So, first, the party responsible for drafting the terms of the contract should be clear about which legal system will be used to interpret the contract. Then, they should still be as clear and specific as possible when actually drafting the warranty clauses.

Furthermore, there are some specific conditions that should be attached to all warranties. First, the warranty period should be specific. It should have specific start and end dates. If the invocation of a warranty repair requires another warranty, then this period should also be defined. Second, a specific remedy should be stated in the contract. This allows all parties to know, up-front, what will and what will not be done. As an example, if a component fails, the contract should be clear about whether it will just be repaired or replaced, or whether it will be redesigned as well, if necessary. Third, conditions which will void the warranty must be clearly stated. Examples include improper maintenance and repairs made by others. Fourth, the contract should provide for a specific time within which the contractor should be notified of a potential warranty defect.

And last, the warranty articles should end with a highlighted sentence stating that the warranties are "exclusive" and "in lieu of all other warranties, whether implied or express." One such clause could be: "FURTHER, IT IS AGREED THAT THE CONTRACTOR MAKES NO OTHER REPRESENTATION OR WARRANTY, EXPRESS OR IMPLIED. THIS WARRANTY AND REMEDIES OUTLINED ABOVE ARE EXCLUSIVE AND IN LIEU OF ALL OTHER WARRANTIES INCLUDING THE WARRANTY OF MERCHANTABILITY AND FITNESS FOR A PARTICULAR PURPOSE."

Remedies available "at law or in equity." Remedies resulting from a dispute that is being decided in the legal system can be granted either "at law" or "in equity." In today's U.S. legal system, the difference is very narrow. "At law" remedies essentially mean those remedies that result in accordance with common law—legally adopted statutes. "In equity" remedies are those that fall outside of the jurisdiction of common law and are ruled on by a judge instead of a jury. In all federal courts and most state courts, civil cases now proceed in the same fashion, regardless of whether they involve legal or equitable redress.

The most important distinction between "at law" and "in equity" is the right to a jury trial in a civil case. Where the plaintiff seeks a remedy of money damages, the plaintiff is entitled to a jury trial, provided the amount sought exceeds an amount specified by statute. Where the plaintiff seeks a remedy that is something other than monetary, the plaintiff is not entitled to a jury trial. Instead, the case is decided by one judge. If a plaintiff asks for both equitable and monetary relief, a jury will be allowed to decide the claims that ask for monetary relief, and a judge will decide the equity claims. Judges are guided by precedent in equity cases, but in the spirit of equity, they have discretion and can rule contrary to precedent.

Precisely for this reason, it is in the best interest of all parties to the contract to specifically exclude remedies available "at law or in equity" that conflict with the provisions of the contract. A contract for power plant construction work, especially new construction work, is usually for large sums of money. The parties to the contract have invested considerable time, effort, and money and therefore normally have structured the contract with very specific remedies in the event of differences. They have done this based on expected returns on these investments and allocation of risk within the contract. If the contract does not exclude remedies at law or in equity that conflict with those spelled out elsewhere in the contract, and if the settlement of differences ends up in court, the intent of the parties for resolving disputes may be ignored by the legal system. This, then, could severely impact the original business model that the contract was structured to protect. To avoid this, the contract should specifically exclude remedies available at law or in equity that conflict with the provisions of the contract and state that the obligations and remedies stated elsewhere in the contract are the only obligations and remedies.

Third-party claims. A third-party claim is a claim by anyone other than the signatories to the contract. It might be by another contractor on-site, or it may be by the family of an employee who was injured on-site. The contract should address what happens in the event of a third-party claim.

A third-party claim usually is designed to recover general damages from the negligent party, which may include pain and suffering, disability, loss of earning capacity, and compensation for future and permanent injuries. And it will be in addition to any workers' compensation payments.

The best way to illustrate possible third-party situations is as follows:

- A worker is injured as a result of negligent conduct by an employee of another subcontractor. For example, a pipefitter is standing on scaffolding erected by the scaffolding contractor. As the pipefitter is performing his work, the scaffolding support cable unexpectedly releases and the scaffold drops. The pipefitter suffers injuries due to this. The pipefitter could have a third-party case against the scaffolding subcontractor, the general contractor, and possibly the owner of the property where the accident occurred. And this would be in addition to any workers' compensation claims. (But note that a worker cannot sue his or her employer for such damages—this is because there is workers' compensation.)

- A worker is injured as a result of defective equipment or tools (ladders, power drills, power saws, etc.) This type of case is referred to as a "third-party products liability," and a lawsuit could be filed against the manufacturer of the defective piece of equipment or tool.

- Another example of a third-party claim would be a truck driver injured in an automobile accident while moving components from the preassembly area to the work area. If this accident was due to the negligence of the automobile driver, who was working for a different contractor, the truck driver would have a third-party case against the other contractor.

Therefore, prudent contracting suggests that the contractor agree to indemnify the owner from third-party claims only to the extent that a claim is due to the negligent action of one of the contractor's employees, during the performance of work and while on the premises. Also, the owner should be required to notify the contractor promptly of the claim, so that the facts can be investigated while the incident is still fresh in the minds of the witnesses. (See the "Indemnification" section later in this chapter for additional information.)

Changes. There are very few power plants that have ever been built where no changes occurred between the time the original contract was signed and the plant was turned over to the owner. Changes are a way of life in the contracting business. Many times, the changes are preplanned, designed, and clearly accepted by all parties. However, there are times when changes are not amicable. Poorly written contract documents are at the heart of many of these. Honest misinterpretations are common. Third-party requirements can lead to changes that neither the owner nor the contractor feel that they should pay for. But unfortunately, there are many contracts where the rules for managing change are not clear, and this leads to disputes, contentious claims, and general ill will among the contracting participants. The potential for cost and schedule consequences of changes must be understood by those directing the change.

For this reason, a method of change management should form one of the cornerstones of the contract document. There should be a clearly written process for handling changes, and this process should be part of the T&Cs section of the contract. As with many of the other basic, or must-have, terms, no single approach to change management will suffice for every contract. Change management should be viewed as a separate project management process, almost independent from the management of the project itself.

Change management can be likened to a three-legged stool. One leg consists of the contract terms. Another is the risk management process, and the third is insurance. In this chapter, we'll limit discussion to the contract language, those terms or clauses that should address, very specifically, how change should be dealt with.

The need for change can come from many directions. It can be owner directed, to satisfy operational parameters of the plant once it is up and running. A classic case would be to add platforms for access to valves that otherwise would require scaffolding for maintenance purposes. Change could come from environmental considerations. Let's say the investors in the plant agree to stricter emissions controls, thereby adding additional air quality control systems.

Change could come from members of the public who want different ingress to and egress from the site, that is, better traffic control. Change could come from a variety of design conditions that were not anticipated when the original plant specifications were developed. Change often comes from conditions encountered by the contractors as they set about building the plant, such as interference of structural steel support columns with ductwork, insufficient clearances between insulation on high-temperature piping and other structures, or equipment anchor bolt holes not aligning with the anchor bolts installed in the foundations due to different drawing revisions. The list goes on and on.

But the heart of the matter is, "How does one manage change when it does arise?" In the best-case scenario, it should be handled using agreed-upon mechanisms that assign the risk for the cost and schedule to those who can best handle it. Therefore, the first step is to define change. Something akin to the following might be used:

> Contractor will be entitled to equitable price and schedule adjustments for changes defined as:
>
> - Owner-initiated or owner-approved changes to the scope of work
> - Delays and/or deficiencies in owner-responsible items
> - Interferences by the owner or by others for whom the owner is responsible
> - Changes in law or permits or other governmental action or inaction
> - Force majeure events as defined elsewhere in the contract

- Subsurface conditions that differ materially from earlier geotechnical investigations
- Suspension or termination of the work not due to the contractor's fault
- Unavailability of craft labor

The next step is to delineate the steps to be used for notifying the responsible parties that a change is forthcoming. This could be done in a number of ways, but at the center of this notification must be timeliness. For example, the change clauses should address how soon the parties are required to notify each other of a pending change. The best way, of course, is instantly. But there are times when this is either not practical or commercially desirable. However, if the contract were structured to require a regular, say, weekly, review meeting of potential, pending, and actual changes, most items would be brought to the attention of all parties in time to formulate action plans that would satisfy everyone.

As an example, if the last 30 minutes of each weekly meeting were set aside for discussions exclusively on changes, the forum would be open. The contract could even stipulate that change requirements that were not addressed at these meetings in an early stage would be disallowed, or they would be treated under the terms of the extra work clauses of the contract, regardless of the cause. One purpose of this is to provide the owner notice while there is still time for the owner to mitigate the pending changes.

But then, there also needs to be a condition of formal notification. Let's say that the contractor knows that a change of access will be required for bringing the preassembled sections of the heat recovery steam generator (HRSG) into position from the preassembly area. Although this may have been discussed at several of the earlier meetings, it should be incumbent upon the contractor to formally notify the owner or general contractor (GC) that in a certain number of days, the contractor will need to have access to a specific side of the unit to bring in the preassembled sections, or else the hydrostatic test date may be in jeopardy. But the key is that this notification should not be some standard one, three, or seven days. It should be notification with enough time to allow the owner or GC to arrange for the proper access without undue interruption of other site activities. And as an incentive to make this timely notification, the contractor should be held responsible financially if the contractor does not do so.

There are many additional aspects of change management that must be addressed somewhere in the contract. For example, one must account for how to price changes (see chapter 2), how to account for changes, how to handle their impact on the schedule, and so forth. These are beyond the scope of this book. But it is worth noting that there is a list of the "Deadly Dozen" issues that can occur on a job site that will almost always lead to changes—changes that must be managed. Read about them in chapter 6.

Delays. Large construction projects frequently incur delays, planned and unplanned. The cost and impact mitigation due to planned delays are usually negotiated between the parties that are affected. The delay is then built into the project, all parties are compensated accordingly, and the job moves on. One example of this could be a major outage, planned for six days a week, 12 hours a day. Part way through the outage, the turbine rotor repair shop notifies the owner that the return of the rotor will be delayed by three weeks. If the turbine work was the project's critical path, then this delay would open the project for a relaxation of schedule along with a reduction of overtime hours. This would be negotiated between the owner and contractors, with the owner expecting a savings of premium time differentials and the contractors expecting an extra for additional time on-site.

Unplanned delays, on the other hand, often occur without time to plan and negotiate the consequences. They may be acts of God. They may be uncontrollable economic conditions. They may be unexpected labor issues, or they may be due to some action, or inaction, by another contractor or owner on-site. The possibility of these occurring on a major power plant construction job needs to be considered in the contract language. For this reason, it is usually prudent to include a delays clause, with specific definitions of what is considered a delay. The clause should state what happens if the delay is or was beyond the control of the owner, beyond the control of the contractor, and/or beyond the control of both. If the delay was beyond the control of both, it is usually considered a force majeure event. The words *force majeure* are French for greater force, which in the contracting world means a force not controllable by any party to the contract.

Force majeure delays are beyond the control of the buyer or seller. Such things as lightning strikes, hurricanes, earthquakes, and the like, commonly called acts of God, are classic cases where no one has control. In such cases, the contract will usually stipulate that the contractor will be allowed a reasonable extension of time; however, since neither party had any control over the event, it is seldom compensable. Other items, outside of acts of God, may also be included, such as labor strikes, war, sabotage, and whatever additional items the parties deem to be included. The important thing to remember is that if it is in doubt, the item should be written into the contract as an exclusionary or force majeure item.

For example, let's assume that the contract includes labor strikes as a force majeure item. Then, let's assume the contractor's personnel go on strike for higher wages. The contractor will claim a force majeure delay, but the client may refuse, saying that the intent of this clause is for causes outside of the control of the parties, and the wage issue is not outside of the contractor's control; the contractor could have given the workers more money. Obviously, more clarification is required.

As part of any delay clause, the time in which the other party is to be notified should be spelled out. If the delay is a force majeure event, the contract should stipulate that the contractor has a fixed number of days to notify the owner of a need for schedule extension. One reason is that the owner may want to take a different course of action than simply extending the schedule. The owner may want to ask the contractor to work additional overtime, for which the owner is willing to pay. Or the owner may want to bring in an additional contractor and split the work between the original contractor and the new one, to recover or avoid lost time.

Applicable law. For contractual T&Cs to be enforceable, they must be tied to some system of enforceable laws. One way to accomplish this is to insert a designated jurisdictional clause similar to the following: "This Agreement is made and shall be construed in accordance with the laws of the United States and the State of New York, and it will govern these terms and conditions, without giving effect to any principles of conflicts of laws. Venue and jurisdiction for all disputes will lie in Westchester County, New York. If for any reason a court of competent jurisdiction finds any provision of this Agreement, or portion thereof, to be unenforceable, that provision of the Agreement will be enforced to the maximum extent permissible so as to affect the intent of the parties, and the remainder of this Agreement will continue in full force and effect."

Dispute resolution. Unfortunately, large, long-term contracts sometimes end with the parties in conflict over some item or issue. This may be because the language of the contract document is not clear, or it may be because the language was never included in the first place. Issues can range from scope, schedule, extra work rates, payments, delays to access, site maintenance, damages, and labor troubles. Whatever the issue, its resolution will be much easier if the framers of the contract document agree on the rules for resolving disputes and then clearly spell them out in the document.

With the U.S. contracting world becoming more and more litigious, owners and contractors increasingly are attempting to avoid the court system to resolve disputes. It is always costly, lengthy, and acrimonious, and it frequently results in a win–lose (and sometimes a lose–lose) result for the parties to the dispute. Often, it also results in the end of any future business relationship. Therefore, many contracts now state that an alternative dispute resolution (ADR) must be attempted first. Such an ADR can be arbitration, either binding or nonbinding. It can be mediation. It can be empowered negotiation, or it can be accomplished through the use of project neutrals who live with the project and can intervene before an issue becomes an irresolvable dispute. Let's look at each of these in light of their applicability to a large power plant construction project.

1. Arbitration. First, let's address a time-honored method that avoids the court system, at least initially: arbitration. Arbitrators act as neutral third parties to hear the evidence and decide the case. Arbitration can be binding or nonbinding. It usually follows a set of procedural rules made by a particular arbitration body. For example, the parties drafting the contract could agree that in the event that a dispute requires arbitration, the arbitration proceeds under the rules of the American Arbitration Association, the International Chamber of Commerce, or many other venues.

Then there is the question of the applicable law. Although the contract may stipulate that the rule of law is the State of New York, the arbitration article could stipulate that the seat of arbitration, and therefore the rule of law, is another jurisdiction.

Nonbinding arbitration is a type of arbitration where the arbitrator makes a determination of the rights of the parties to the dispute, but this determination is not binding upon them and no enforceable arbitration award is issued. The award is in effect an advisory opinion of the arbitrator's view of the respective merits of the parties' cases. Nonbinding arbitration is used in connection with attempts to reach a negotiated settlement. The role of an arbitrator in nonbinding arbitration is, on the surface, similar to that of a mediator in a mediation. However, the principal distinction is that whereas a mediator will try to help the parties find a middle ground on which to compromise, the arbitrator remains totally removed from the settlement process and will only give a determination of liability and, if appropriate, an indication of the quantum of damages payable. Read on for more on mediation.

Subsequent to a nonbinding arbitration, the parties remain free to pursue their claims either through the courts or by way of a binding arbitration, although in practice a settlement is the most common outcome. The award and reasoning in a nonbinding arbitration is almost invariably inadmissible in any subsequent action in the courts or in another arbitration tribunal.

2. Mediation. An alternative to arbitration is mediation. Here, the primary objective of the mediated settlement is a fair–fair solution and return to business as usual as quickly as possible. There are times when it is very appropriate to mediate, and there are times when it is not. The key is to understand the difference.

- Do mediate when:
  - The parties seek an end to the problem, not the relationship.
  - The dispute is to be kept private.
  - The law cannot provide the remedy sought.
  - The parties want to minimize work disruptions.
  - A party wants to avoid an adverse precedent.
- Do not mediate when:
  - One party is unwilling to mediate.
  - One party wants to go for the jackpot.
  - One party wants to establish a precedent.
  - The dispute involves a crime.

When mediation is the chosen course of dispute resolution, it is important that a skilled mediator is chosen. Not only should the mediator be trained as a mediator, he or she should also be an expert in the field of construction, preferably in large power or at least industrial plant construction projects. The key to avoiding controversy over who the mediator will be is to select several candidates at the outset of the project, during the contract negotiating stage, and then write their names into the contract.

3. Empowered negotiating. Going outside of arbitration or mediation, another form of dispute resolution is also used, called empowered negotiating. *Empowered negotiating* starts with the same preparation of analysis and calculations of damages that one would do for arbitration or mediation. However, rather than preparing fully developed claim documents with supporting exhibits, a lesser level of effort is applied. The idea is to spend only a modest sum of money on claim preparation, with the resultant expectation of receiving something less than 100% of the claim. This method can be an attractive alternative to spending considerably more in document preparation for the rather low probability of recovering significantly more dollars, while running the risk of being shut out completely.

4. Project neutrals. Finally, in an attempt to get in front of potential disputes, some owners and contractors are pushing dispute resolution into the early stages of projects by assigning an ADR-trained person to the project to follow it from groundbreaking to completion. The key to success here is the concept that this person has only one client, the project itself; hence the term *project neutral*. This person acts as an impartial mediator, proactively inserting him- or herself into disputes long before they become contentious. In the words Kenneth C. Gibbs, a mediator/arbitrator with JAMS Global Engineering and Construction Group, one of the nation's largest private providers of ADR services:

> I have had the opportunity to serve as the project neutral on many major projects. Using a proactive approach, I work with the project teams to look ahead and avoid disputes altogether by identifying and addressing potential problems before they become real issues. The key is always in being able to bring the focus of the issue to the business side instead of focusing on the legalities. In this manner, the parties to the project can deal with what they understand best, the dollars and cents of the issues, not their legal nuances. The project neutral concept takes the benefits of mediation and applies them in a cost-effective manner to construction management.

Indemnification. Construction contracts are intentionally drafted to allocate risk between the parties. An indemnity clause is a contractual device to shift common law and statutory risk associated with a party's negligent acts from one party to another. In the construction industry, the owner wants to allocate the responsibility for negligent acts from itself to the design professionals and the contractor. In turn, the contractor wants to shift that risk to the subcontractors and suppliers. All parties usually carry insurance to cover themselves for some of this risk, but there is a cost associated with this insurance coverage.

To complicate matters, insurance carriers inject themselves into the contracting process and require their insured clients to limit their exposure through the use of "express indemnity" clauses. They want to ensure that the limits of indemnification and the associated risks are clearly defined. Let's look at some examples.

> **Example 1.** A subcontractor installed a high-pressure steam valve manufactured by a third party. The valve failed, and live steam injured several employees and caused some property damage. The indemnification between the subcontractor and the GC required the subcontractor to indemnify the GC for loss which was in *any way* connected with the subcontractor's work. While during the trial the court agreed that the subcontractor did not install the valve negligently, it still ruled in favor of the GC. Why? Because the contract language explicitly stated that the subcontractor was to indemnify the GC for losses that were *in any way* connected with its work. The indemnification clause did not exclude acts of nonnegligence.

> **Example 2.** An insulating contractor enters into a contract with a boiler contractor to insulate the boiler upon completion of the mechanical boiler work. The insulating contractor agrees to indemnify and hold the boiler contractor harmless from any and all claims or damages arising out of the performance of the contract, whether caused by the insulating contractor's own negligence, the boiler contractor's negligence, or any of the boiler contractor's employees' negligence. Further, the insulator expressly releases the boiler contractor and waives all rights of action against the boiler contractor for any such claims or damages.

During the course of the insulating work, the boiler contractor agrees to provide the insulator with scaffolding for the insulator's use. The scaffolding falls due to improper modifications made to it by the boiler contractor's employees, and an employee of the insulting contractor is injured. The employee collects workers' compensation, and the insurance carrier successfully sues the boiler contractor for negligence to recover the benefits it paid to the employee. The boiler contractor then sues the insulating contractor for reimbursement of these damages, citing the insulator's duty to indemnify it for any claims arising out of the contract, even if they resulted from the boiler contractor's negligence. The suit is successful!

Why? As before, because the contract specifically stated that this was the intent of the contracting parties. These issues often arise when one contractor, normally the lower-tier contractor, is much smaller than the upper-tier contractor or owner. First, the smaller contractor often does not read the fine print of the contract. But even when the smaller contractor does, the theory is (a) nothing will go wrong; (b) if something goes wrong, everyone will share in proportion to their culpability; (c) if something goes wrong and the sharing does not happen, the smaller contractor will just go out of business—maybe returning the next month with a different contracting license. Also, remember that just because one party is awarded damages from the other, this does not necessarily mean the other will, or can pay, especially if the proper insurance policies are not in place.

When negotiating contracts, the risk of any action should be carried by the party best suited to manage it. This is especially true for indemnification. Why should the "little guy" indemnify the "big guy?" The ultimate cost to the contract, and the project, is higher due to the higher insurance premiums that the lower-tier subcontractor will incur. The parties to the contract should take care that the language in their indemnity agreements properly states the scope of indemnification intended.

Therefore, it normally behooves both parties to restrict indemnity obligations to being the responsibility of the negligent party, and even more so, that it will be due to its performance of the work while on the job site. In other words, it would not cover an accident caused during an employee's drive home from a bar where the employee had stopped after work to "discuss" tomorrow's work schedule with some coworkers. (Also, see the third-party clause earlier in this chapter).

Proprietary and commercial property protection. Most parties to the contract will own some kind of documents or other items that they wish to keep others from duplicating and/or using for commercial gain, without reciprocal compensation. These items may be shop or other detail drawings and designs. They may be specially developed software products or a host of other patented or copyrighted materials or processes. The contract should clearly spell out that the use of these items, outside of the confines of the contract work, is prohibited. A typical case is the use of confidential drawings, especially drawings from an OEM.

Often, a plant hires the OEM to install or replace/repair equipment and components because of the OEM's ownership or access to OEM *confidential* drawings and designs. This ensures that the work will be done to the specifications that the OEM intended when the part or equipment was designed and that it will perform accordingly. However, this usually comes at a price—the OEM may charge a higher price than a third-party contractor because the OEM has invested considerable time and money into developing and proving the part or equipment, and the OEM is now guaranteeing its performance. Since third-party contractors do not have these up-front costs to recover, they may be able to provide the same work for a lower cost if only they had access to the same drawings. So the OEM will ask that the contract prohibit the owner from using these drawings or designs without the OEM's explicit approval. In other words, the OEM does not want its competitors to have access to its proprietary information and then be the lower bidder on the next job.

Performance guarantees. In addition to guaranteeing project completion by a certain date, many larger power plant construction projects also include certain performance obligations. These obligations are usually related to equipment provided by the contractor that forms part of the final plant operation, such as valves, motors, pumps, and instruments. To avoid disputes, a win–win contract document will clearly spell out the parameters associated with guaranteeing performance. First, guarantee points should be clearly stated. Second, the methods and means for the testing itself should be written. Third, it should be clarified whether the testing is to be performed by the contractor, the owner, or a third party.

But not only should the guarantee points and performance methods be spelled out, but also the existing conditions on which the guarantee points are dependent must be included. Finally, specific remedies should be included in the event that the performance guarantees are not met. These remedies may be as simple as replacing the item (e.g., the controls of a feedwater control valve). But in the event of an inaccessible part (e.g., some boiler bank tubes in the center of the generating bank between the upper and lower steam drums), the replacement may not be feasible, and payment may have to be adjusted to compensate for the resultant reduced throughput.

All of the above must-have contract clauses are those that should be included in any major power plant construction contract. They are the ones without which any or all parties to the contract could be open to major risk. Again, it is all about who carries what risk.

Nice-to-have

But there are also a series of contract T&Cs without which a contract can still be administered. These are called nice-to-haves. Their definitions change depending on the contract and the needs of the contracting parties.

- Default. What is the definition of contractor default, and what remedies are available to the owner?
- Suspension. What constitutes contract suspension, and what remedies are available to the contractor?
- Termination. What constitutes contract termination, and what remedies are available to the terminated party?
- Conflicts of contract provisions. Which contract documents take precedence?
- Time-is-of-the-essence. The general rule is that time is not of the essence unless the contract expressly so provides.
- Nonassignment. What are the restrictions on assigning the work to another entity?
- Extra work. What are the definitions of and rules and costs for work outside the base scope?
- Noncollusion. Has there been any action in restraint of free competitive bidding?

- Escalation. What are the rules for passing on increased costs of labor, equipment and materials?

Who Is Protecting from Whom?

The manner in which all of the above T&Cs, both must-haves and nice-to-haves, are addressed differs significantly depending on who is trying to protect themselves from whom. A plant owner will be looking for as much shifting of risk to the GC as possible. On the other hand, the GC will be looking to shed as much risk back to the owner as possible, and when negotiating with the subcontractors, the GC will be looking to shift risk down to them as well. Meanwhile, the subcontractor will be looking to shed risk by shifting it to back to the GC. The next two figures compare some of the more salient T&Cs from the viewpoints of these different parties. Figure 5–1 shows the relationship between the owner and the GC, and figure 5–2 shows the relationship between the GC and the subcontractor.

TERMS AND CONDITIONS
Owner — General Contractor Contract

Owner's View

INCLUDE	EXCLUDE
Indemnification clauses	Limited or qualified indemnification
A/E's decision as final and binding	Limited A/E's decision clauses
Limiting extra work clauses	Open-ended extra work clauses
No damages for delay or disruption	Clauses permitting delay claims
Liquidated damages clauses for delayed completion	Attempts to qualify or limit liquidated damages
Payment clauses requiring release of liens and claims	
Termination for cause clause limiting exposure to claims	
Termination for convenience clauses limiting exposure to claims	

General Contractor's View

INCLUDE	EXCLUDE
Limited indemnity only for negligence	Payment contingent on events other than performance of the work
Concealed conditions clause	A/E decision being final
Recovery for extra work	Detailed and complex extra work clauses
Recovery of delay related damages	No damages for delay clauses
Recovery of overhead and profit in the event of termination	Liquidated damages clauses

Fig. 5–1. Owner and GC contract clauses

TERMS AND CONDITIONS
General Contractor – Subcontractor Contract

General Contractor's View

INCLUDE	EXCLUDE
Incorporate by reference all contract documents	Clauses that bind subcontractor only to technical specifications
Pay when paid clauses	Limited indemnity clauses
No damages for delay clauses	Clauses allowing for recovery of damages or termination in the event of delays
Clauses allowing for recovery of overhead and profit on deductive change orders	
Termination for convenience with payment of only actual loss of work supplied	
Termination for cause upon two days written notice	

Subcontractor's View

INCLUDE	EXCLUDE
Clause limiting scope only to plans and technical specifications	Incorporation by reference clauses
Unconditional payment clauses	Pay when paid clauses
Basic extra work clauses	Complex extra work clauses
Clauses entitling subcontractor to delay damages	No damages for delay clauses
Clauses permitting termination for nonpayment	

Fig. 5–2. GC and subcontractor contract clauses

First, let's look at figure 5–1, a comparison of the contract clauses wish list from the owner's point of view, as opposed to the GC's point of view. Owners want broad indemnifications. They want to include protection against third-party claims, whereas contractors want to limit these protections to their own negligence. Owners want to give their architectural engineers the latitude to ensure that the work meets the intent of the specifications without granting additional compensation or time to the contractor. Owners want to minimize extra work, whereas contractors want to be paid for anything even resembling out-of-scope efforts. Contractors also want to be paid for disruptions due to concealed conditions, such as finding underground piping or electrical cables, of which they were not advised. Owners want to shift the cost of damages due to delays to contractors by increasing the liquidated damages; contractors want the exact opposite. Contractors want to reduce their exposure to liquidated damages and want to qualify specifically how these damages are triggered.

In the event that the owner terminates the contract, the contractor wants to recover the overhead and profit the contractor had planned for the total project, while the owner wants to limit this to only that portion directly associated with the work performed.

Figure 5–2 shows a similar comparison of contract clauses, but this time it is a comparison between what the GC wants versus what the subcontractor wants. Specifically, the GC is looking to shift all scope risk to the sub by trying to incorporate any and all documents related to the project into the contract between the GC and the sub. The GC is also looking to stay cash neutral by asking for a "pay when paid" clause where the GC does not have to pay the sub until the GC has been paid. This can have very serious consequences for the sub, since in the construction world, cash is king. Therefore, the subcontractor should ask for a clause permitting the sub to terminate the contract for nonpayment.

Since the subcontractor is at least one step removed from the owner, and usually not in a position to reach the owner contractually, the sub must be cognizant of issues that may be created by the owner, flow through the GC, and then affect the sub directly. "Pay when paid" is one such issue. But so are delays. The delays may not be the fault of the GC; they may be the fault of the owner or even another contractor on-site. So the sub wants the contract language to protect the sub in the event that there are costs due to delays by others, beyond the GC.

Summary

The commercial side of the contract should be the least used part of the contract document. In a perfect world, there would be no delays, no scope changes, and therefore no disputes. On smaller, faster jobs, that is often the case. However, in the real world, when projects are large and complex, it is impractical to expect perfection. So the T&Cs of the contract do have their place.

But to be helpful in resolving the issues that do arise, the T&Cs have to be practical. They should not be one-sided, they should be clear, and they should outline a resolution process that is acceptable to all parties. They have to be legally sound and usable in the jurisdiction where the issues will be resolved.

All parties to the contract must understand that they have different levels of responsibilities to each other. The owner has a responsibility to the investors of the plant, the architectural engineer, and the GC, who in turn is pulled between the owner and the subcontractors. The subcontractors may be impacted by others against whom they have no recourse, yet they still have a responsibility to the GC. The win–win contract will take into consideration the needs of all these parties and strive to protect each one. As one GC stated: "I'd rather have a tough contract that is easily administered than a weak contract that cannot be administered."

But in the end, or actually at the beginning, it is crucial to heed the warning that trying to be your own lawyer is a foolish endeavor. As stated at the beginning of this chapter, contract legalese requires careful preparation. If there ever arises an occasion where the T&Cs of a contract are required to resolve an issue, they have to be very clear. They have to be enforceable in a court of law. And they have to provide all of the protections the company requires, within the confines of the laws in the jurisdiction where they are being used. This means that expert help should always be used in their preparation. Trying to go it alone when preparing a contract is a risky effort fraught with the potential of disastrous consequences.

Risk Management 6
(with content from Mark Bridgers)

Up to this point, we have focused mostly on pre-site activities. We worked our way through the preplanning stages of contracting, we went into details about the actual planning once a decision was made to proceed with the project, and we looked at how to put the contract together, from a numbers perspective and from a terms and conditions (T&Cs) position. We also touched on risk management. For example, we addressed risks that needed to be assessed before making the decision to proceed with the project. We looked at ways of mitigation using a lessons learned approach. We looked at ways of risk mitigation by exploring different project delivery systems. We went through scenarios of contingency planning and managing.

We described how the bidding and budgeting process must be managed, much as a project in and of itself. We addressed how to write or interpret a specification where the norm is to shift risk to the other party. We went through scenarios of defining scope, schedule, pricing, and payments, and we calculated the value, or lack thereof, of penalties and bonuses.

We reviewed codes, standards, and regulations and the necessity of ensuring that they are correctly followed to avoid being trapped by their requirements. We looked at sourcing personnel, vendors, equipment, tools, and consumables. We addressed the importance of planning for these resources to avoid unpleasant surprises at the last minute. And we discussed T&Cs from a perspective of risk mitigation.

The long history and experience is that the failure rate for construction projects is high. Why haven't advances in project management science, computer technology, and communications been more effectively brought to bear in this business? Because of their complexity and because of the inherent instability created by the contractual structure, which contains incentives and disincentives to proactively solve problems and seek to avoid blame—see the section below on the "black swan" effect.

But risk management is important because at the beginning of a project, team members are optimistic and believe that best practices in project management will minimize the chance of failure. In reality, success is not predictable or guaranteed in the real world. Good planning is not necessarily the use of the most optimistic date for every work element of a three-year project. A more rational approach may be to use a Monte Carlo method of random analysis to bound the time frames expected for the project's tasks and then produce a schedule of the most likely durations and completion dates, given the most recent project histories for that environment. (Monte Carlo analysis is not covered in this book.)

It is important to stay alert to the arrival of complexity by checking to see if established project management metrics are reliable and if critical activities are increasing in number on the Critical Path Method (CPM) schedule or represent more than 50% of the activities. It is also important to note if CPM updates are overshadowed or made outdated by events.[1]

That is to say "the best laid plans of mice and men often go astray." We know that reality will set in, and no matter how carefully and thoughtfully we developed our plan, and no matter how craftily we structured our contracts, certain things just *will not happen as planned*. So we come to this chapter on risk management.

Risk management can be likened to a three-legged stool (fig. 6–1). One leg represents the contract and its language, that is, the T&Cs discussed in chapter 5. The second leg represents the claims management process: It must be clarified that not all claims are contentious, but they can still impart a risk to the cost, schedule, or quality of the project. Finally, the third leg represents the project insurance protections. Without any one of these three legs, the risk management process is subject to failure.

As noted, the leg representing contract language is addressed in chapter 5. In this chapter, we will address the remaining two legs, the claims management process and the insurance protections. Since both of these are part of the risk management process, they must be understood by those preparing the construction contracts as well as by the site management team that follows. Numerous books have been written about each subject, as well as risk management itself. This chapter does not cover these subjects in detail; it only addresses them from the view of the site management team because the intent is to familiarize the team members with the steps they must take to protect their position when risk issues arise. But before we address these last two legs, we first must understand what constitutes risk.

Fig. 6-1. Risk management stool (Courtesy of Construction Business Associates LLC)

Risk Identification

So what is risk as it relates to building or rebuilding a power plant? It can be thought of as a factor, thing, element, or course involving uncertainty to the conclusion of the success factors of the project, namely safety, quality, schedule, or cost. For example, risks to safety could be anything from inadequate training to physical fatigue due to long work hours, inadequate safety equipment to poor traffic control, and not following prescribed procedures for various activities such as heavy equipment lifts. Inclement weather could be considered a risk to schedule. Importation of poorly made foreign materials could be considered a risk to quality. And estimating and/or budget accuracy could be considered a risk to cost. Note that risk, as discussed here, does not include what many consider as the risk (or chance) of opportunities; only risk that may be detrimental to a project is discussed herein.

To manage risk, typically a risk identification or assessment log is developed that lists particular risks or risk types and describes their basic characteristics. The identification process can be as simple as a single person writing down risks observed on a previous project or as complex as hiring a professional risk management expert to work with the project team to develop a very detailed risk log. Some organizations have full-time risk assessment specialists on staff to assist project teams in their risk mitigation efforts.

Once a list of risks is developed, sometimes called Phase 1—Identify the Risks, the risks must be assessed as to their impact on the project in the event they actually were to be encountered. In figure 6–2 this is shown as Phase 2—Analyze. The risks are evaluated based on how they might impact project safety, quality, schedule, or cost. This evaluation can be prepared using a project severity index, say, 0 through 10. In this example, 0 is equal to no impact and 10 is extremely serious. Then, multiplying these numbers together (and excluding 0 since multiplying by 0 would result in a zero value), the result would determine the maximum overall impact on the project if the risks were to become reality. Some assessors even color-code those risks with the highest impact as red, the intermediate ones as yellow, and those that are more or less normal as green.

Note that this result, at this point of the risk assessment, does not take into account the probability of the risk occurring. It just shows the potential impact if the risk actually materializes. Many risk assessors include an additional factor called probability. This is shown in figure 6–2, but is not used in the calculations. Essentially, the previous risk assessment calculation would be multiplied by a subjective probability of occurrence, and then this new total would be used to prioritize the various risks identified. In the example in figure 6–2, these totals would be 81, 270, 162, and 168. As one can see, this would move the risk currently ranked as number 1 to number 4.

This second step, the probability analysis, is a dangerous risk mitigation tool. By adding this additional step to the risk analysis process, one *assumes* what may or may not occur, based on very little but hunches and guesses. It completely ignores what more and more experts in the risk probability field refer to as the "black swan effect." This can lead to a false sense of security because if one of these risks becomes a black swan incident, and if it was ranked low in the list due to a low probability of occurrence (which is one of the black swan parameters), the results could be catastrophic. (For more on this, please read on a bit further in this chapter.)

Risk # or Rank	Phase 1: Identify the Risks	Phase 2: Analyze						Phase 3: Mitigate the Risks		Phase 4: Update
	Potential Risk	Safety	Quality	Schedule	Cost	Probability*	TOTAL	Mitigating Actions to be Taken	Assigned Person and Date to Mitigate	Comments / Status
1	Inadequate trench slope in the event of heavy rains.	9	0	3	3	0	81	Increase trenching slope per OSHA requirements.	Project Engineer. 5/12/2013	None to date.
2	Row 1 static seals were found damaged during the last unit inspection (both units). Only 6 seals are available in the LTP parts kits and more may be needed.	0	0	9	6	5	54	Check inventory in warehouse. Provide up-front part ordering information and list.	TG Technician. 3/15/2013	Ordered more seals. Arrival expected two weeks before outage commences.
3	Possibility of discovering unexpected tube thinning during inspection	0	0	9	3	6	27	Check inventory in warehouse. Provide up-front part ordering information and list.	Boiler Technician. 3/15/2013	1,000 feet of spare tubung available in warehouse.
4	Risk of contact with lizards, spiders, snakes, etc., due to remote location of the plant	6	0	2	2	7	24	Advise employees of the potential for contact with wildlife during the walk up to the Plant. Advise no contact with wildlife. Heighten awareness during safety orientations	Site Safety Coordinator. Every morning meeting.	Signs posted, in English and Spanish.
etc.										
etc.										

Fig. 6-2. Risk assessment log. Note that probability is *not* used for calculating the total. (Courtesy of Construction Business Associates LLC)

After analyzing the impact a risk could have on a project (with or without the use of a probability calculation), a mitigation plan must be developed. This is shown as the third phase in figure 6–2. Actions to be taken are described and assignments are made as to who has the responsibility to follow through and a date is set as to when. Finally, Phase 4 is an update status that is used to ensure that risk potentials are acted upon in a timely manner.

There are many ways to develop a risk identification or assessment log. The important point is that such a log must be prepared. The example in this chapter happens to be a snippet from a log prepared for a major power plant turbine and boiler overhaul coinciding with a scrubber tie-in somewhere in the southwest United States. One can envision how many pages might be written for a greenfield new plant project and see the many days of effort this can entail. The sooner this effort begins, the more mitigating its effects can be.

The Black Swan Effect

As discussed above, ranking the identified risks in the risk assessment log can be as simple as multiplying the perceived impacts to safety, quality, schedule, and cost and then using these results to determine a severity listing. This is what is represented in figure 6–2. Or, one can take the additional step of assigning a probability number and rearranging the severity listing accordingly. However, when one does this, there is an inherent danger, often referred to as the black swan effect. This effect comes about from assigning a low-probability number to a potential risk because it has never, or very seldom, occurred on any other project. Now, if in addition to a low probability of occurrence, it also has a low impact on the project if it were to occur, then that potential risk *should be* low ranked—as it would have been even without the probability factor. However, if the potential risk would have a major impact on the project if it were to materialize, then it should be placed very high in the severity listing, probability notwithstanding.

So where did this "black swan" phrase come from? Some centuries ago, Europeans had never seen any swans other than white swans. These were the only swans that existed in Europe at that time. But then, some of these people traveled the world and discovered Australia. Lo and behold, in Australia there were nonwhite swans. At first, these discoverers did not even consider these animals as swans, since their perception of a swan was that it had to be white. But as time moved on, some of these swans were brought to Europe. However, they were in such a minority, that they were considered a rare animal, an outlier if you will.

Today, a black swan event refers to an event that is very likely not to occur, but it now has the additional parameter that if it does occur, it could be catastrophic. For example, think of a major hurricane striking a populated city, or think of an offshore oil rig blowout preventer failing and spilling millions of gallons of oil into the sea. In the power plant construction industry, similar black swan events also occur.

Capital construction projects, such as a new plant build, and even major outages and other power plant construction activities do not lend themselves to the type of risk management and analysis that are generally used on more traditional projects. In the traditional risk assessment process, the probability of risks occurring is based on a bell-shaped curve often referred to as a normal distribution pattern.

This curve is formed by taking a large sampling of many projects over a long period of time and plotting the frequency of occurrence of any particular risk across all of the sampled projects. Doing this for all of the potential risks associated with the project at hand, one can then determine if each of the risks falls within the normal distribution, or bell curve, or at the fringes of it.

Then, if the risk is within the bell curve's 99.8% left-to-right boundary, the risk should be considered. This is known as being within the third standard deviation. If it lies beyond this parameter, the risk is considered an outlier and has an extremely low probability of ever occurring.

Unfortunately, if this low-probability risk *does* occur, it is frequently associated with major consequences. One example of a low-probability, high-impact risk is the cost of raw materials used in the building of a major power plant. Between the years 1989 and 2003, the trading price of copper, such as that used in the manufacture of many power plant components, as well as all of the electrical cabling and wiring, averaged just under $1 per pound (in U.S. dollars). However, by 2011 it had climbed by a factor of almost five, to $4.50 per pound. As of this writing, it is hovering around $3.25 per pound. So if a budget were developed in the early 2000s, the estimator would certainly have assumed a copper price of around $1.00, or maybe $1.50 maximum; that was the historic trend. But if that budget was not revisited until contract pricing was being solicited, say, early in the year 2008 when copper prices had quadrupled to $4.00 per pound, the budget would have been significantly overrun.

Similar cases have occurred with the price of steel and cement. Also, labor pricing is not a fixed known. Contractor and subcontractor bankruptcies fall into this risk pool. And then there may actually be real catastrophes such as storm-related flooding, especially in the basement of an existing power plant where all of the power and control cables for the total plant are located and major revamp work is going on. Or what about a major fire, right during the time when the schedule is tight and every hour is critical? These are often referred to as black swan events.

Because of these risks, claims and insurance are often used as mitigation and recovery tools. Read on for a further discussion on both of these subjects.

Claims Management

As discussed above, risk management can be likened to a three-legged stool where one of the legs is claims management. This is because the use of claims as a tool to recover from unexpected costs due to unforeseen events is a common risk mitigation approach. Periodically, this approach is even encouraged by owners and contractors as a tool to document changes in schedule, scope, and sometimes quality. However, most of the time claims, especially from contractor to the general contractor (GC) or owner, become contentious. Therefore the focus of the following discussion is on avoiding claims in the first place.

In the early days of power plant construction, in the days of regulated utilities, claims avoidance and its parallel risk management process were not at the forefront. The emphasis was on project completion and reliability of the plant. Today, with the emphasis on economics, this has changed. With the tighter economic environment that many contractors, architectural engineering (AE) firms, GCs, and owners are experiencing, profits are being squeezed and the participants in the projects fight harder for every dollar they believe is theirs. This leads to tougher negotiations, tighter contract terms, and stricter enforcement of contract language, especially on larger projects (>$100 million).

For these reasons, it is important today that power plant construction projects are structured to minimize claims, and that they are structured to expedite disputes. This is not easy. Managing risk has become increasingly complex. Technological innovation, globalization, and increased accountability at the senior management level of the corporate world have changed how risk must be managed today. Add that to the shortage of personnel skilled in managing power plant projects, and we have a situation that requires careful planning, contracting, and site execution—from the outset of the project—to avoid claims.

There are many steps that owners, designers, and contractors can take toward ensuring that their projects are completed on time, within budget, and without claims and litigation. First, learn all you can about your partner. Risk shifting by owners, AEs, and GCs has become the norm. Large firms often create special corporations to build a project, with the specific intent of limiting their liabilities. Their purpose is to shelter the parent company from legal problems and financial obligations in the event of issues down the road. It may appear that corporate money is paying for the project, but the special corporation is often financed by parties with no connections to the corporate firm. To make matters worse, these financiers often have no money available for cost overruns, which means that if contractors get into trouble through no fault of their own, they may have no one to turn to.

After vetting the partners in the transaction, and knowing their financial strengths and weaknesses, good contracting must be enacted. The T&Cs of the contract are often bigger risks than timely or efficient contract performance. For example, force majeure used to be a protection for the contractor. In the event of forces beyond their control, contractors were allowed compensatory time and sometimes were even paid for costs incurred. Today some owners are denying this protection and forcing acceleration without compensation. Contractors used to expect protection when encountering site conditions different from what was expected; today, this risk is frequently shifted back to the contractors. For reasons such as these, it becomes increasingly important for all parties to the contract to clearly understand their obligations to each other and to third parties as well.

In chapter 5, there is a list of specific must-have contractual clauses that should be in every contract and understood by all site management personnel. Some of these clauses impose specific obligations on the parties that, if not met, may cause the offended party to lose its right of redress.

For example, most delay and force majeure clauses require notification within a certain time after the occurrence of the event. Suppose a hurricane shuts down the job for a few days. If the contract requires the owner to be notified within three days of the event, but if the contractor waits until the end of the job to claim for time, the owner may have the right to deny the claim, even though the owner knew there was a hurricane. Therefore, it is necessary for the parties to understand their obligations in order to preserve their rights of future redress.

In today's litigious world, a strong step toward claims avoidance would be a mandatory process of review and negotiations that has to be followed before any legal actions are started. The first such step could be a systematic process of reviewing potential claims. For large new construction projects and overhauls with unknown scopes, the participants could be required to hold weekly meetings to review any items of possible contention, sometimes referred to as "issues bubbling beneath the surface." (For example, what work will be required once the turbine has been opened and inspected?) The idea is that if these items are brought to the attention of the parties at the earliest stages of discovery, then often there is time for work-arounds that will lessen or eliminate problems down the road. Even if the contract documents are silent on this issue, site management teams still can implement their own review procedures to facilitate the avoidance of claims.

The next requirement may be a mechanism stipulated in the contract that encourages both sides to meet and discuss problematic items at varying levels of management. This kind of provision can greatly enhance the resolution of claims by negotiation among senior managers of both parties, eliminating the need for a legal resolution.

As can be seen, most of the foregoing can be summed up in one word: *communication*. Communication is the key to many of the issues that lead to claims. Figure 6–3 shows a classic case of the lack of communication during a construction project. Unfortunately, owners do not always get what they want, and contractors do not always build what was designed. Construction requires very extensive communication efforts by all parties. Site visits, models, renderings, and computer simulations can help explain what is intended more clearly than just a set of plans and specifications.

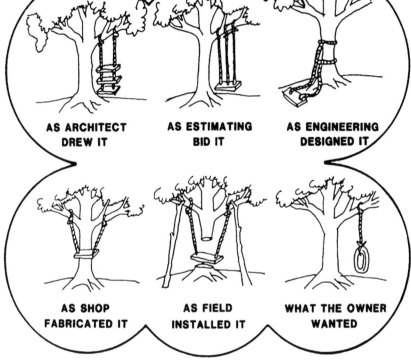

Fig. 6–3. Claims can be avoided with proper communication. (Courtesy of WIlson Management, Inc.)

On complex new construction projects (and even for major retrofits where piping or various pieces of equipment must be shoe-horned into position), computer simulations can greatly enhance the understanding of the task at hand. This can happen both in the office during the design stage and in the field before the installation work proceeds. Some years ago, this type of modeling was very expensive, and only the largest contractors could afford the investment. Today, this is no longer the case. This kind of simulation can quickly highlight problem areas in time to develop alternate approaches or build a clear case for a contract adjustment. No longer can designers and contractors use the excuse of "unforeseen obstructions or interferences" as effectively as they could in the past. The tools now exist to anticipate and develop mitigation strategies for these issues and thereby avoid contentious claims.

Being prepared

A major step toward claims avoidance is to be prepared in the first place. Being prepared forces one to be cognizant of the potential for claims, whether as the claimant or the one defending against the claim. The first step is maintaining all documentation. It is not sufficient to only keep the *latest* version of a document. For the resolution of most claims, a trail of evidence, a *history*, must be available. This suggests that all versions and revisions of pertinent documentation must be available and must be able to be linked to each other and to the latest version, at least up to the date of the claim. Otherwise, the opposing party may reconstruct the evidence, and surely their version will not be your version. The most obvious document to which this applies is the construction schedule.

For example, let's assume that the GC is the author and keeper of the overall construction schedule (as it should be!). The updated schedule is issued weekly, based on input from all of the subcontractors. Now suppose that the project is nearing completion, and the electrical contractor has not yet set the motor control center or pulled the power cables to the boiler feedwater pumps, although the original schedule showed this being complete by now. In actuality, however, as the weeks of the project progressed, the work in this area continued to be delayed due to lack of access and problems in constructing the motor control center enclosure, none of which were under the electrical contractor's control. Now comes the time for filling the boiler with water for the hydrotest, and the feedwater pumps are not operational. The boiler contractor claims for an extension of time and related delay charges to the GC, who in turn claims against the electrical contractor. If the electrical contractor did not keep the entire series of schedules as they were developed and issued, and especially if there was no other documentation contesting the delays, this contractor will have a difficult time proving that others caused the delay. It is extremely important to keep *everything*!

In addition to keeping all versions of the schedule, it is also important to keep the original estimate and all of its variations. This can be very useful when arguing against a claim or when making a claim to show that the item in dispute was or was not different when originally bid. The same goes for documentation of the site visit and any information provided by the owner and GC at that time (see Appendix D, "Job-site Visit Information Sheet"). It is important to file and have readily available *all* documents until the project is completed and all of the paperwork, changes, and disputes have been resolved, sometimes several years beyond completion of the site work. In fact, many companies have specific retention requirements for all documentation, and they make it a part of their quality control program.

The claims process

At the outset of the job, the potential for claims must be considered, whether from the viewpoint of the claimant or the one defending against the claim. The modern power plant has yet to be built where there were no claims or there was not at least a thought given to the making of a claim. As stated earlier, that does not mean that claims have to be contentious. Many times, claims are clearly legitimate and readily accepted. The following is not intended to be a comprehensive treatise on claims processing; rather, it is intended to be a primer on what creates claims and what steps are required to be in a position to manage a claim in the event that one is made. Ideally, if all involved are knowledgeable about these steps, contentious claims can be avoided.

There are several steps required to successfully formulate and present a claim for additional costs. First, it must be recognized that something on the job is going wrong. This is not a guarantee that anyone is due more money, but it may be an indication that compensation is warranted. Second, it must be determined what is outside of the contract scope and the costs involved. Third, one party must convince the other that there are merits to the case and the associated costs.

But how does one provide the proof of cost impact or defend against it? Very simply, set up cost codes at the outset of the job, and then diligently input the data on a regular basis. At the time there is even a hint of a potential claim, each party should compare the related costs to actual input data and verify that they are representative of the potential claim. That, in itself, does not constitute acceptance of the claim nor does it constitute agreement that the costs are exact and final, but it does set up the basis from which to negotiate any resulting claim.

Schedule data can be equally important. As shown in the earlier example of the unavailability of the boiler feedwater pump, the contractor's schedule should clearly demonstrate how the work was intended to be completed—the "as-sold" schedule. The work sequence and time durations should be actual intended ones, not simply something to satisfy the contract requirements or use up all available time. It should be sufficiently detailed to clearly show the use of major equipment, movement of materials, number of crews (manpower loaded), and possibly cost loaded (expected cash flow in and out). The schedule should be updated on a regular basis throughout the life of the project and distributed to all parties. You do not want to find yourself in the scenario shown in figure 6–4 and then try to defend a claim.

Fig. 6–4. A detailed and updated schedule is key for a successful project. (Courtesy of Wilson Management, Inc.)

Next, all claims must be documented, and that requires a good record-keeping system. It is therefore very important that each party to the contract takes a hard look at their record-keeping system, runs some tests on it, and sees how well it performs. If it does not meet expectations, then it needs to be modified.

Let's take the case of a major plant overhaul, one that has been planned and scheduled for three or four years. The scope is all encompassing, from turbine reblading to boiler tube replacement to pump and valve replacement/repacking and more. The job is awarded to a major contractor who breaks it into smaller packages for subcontracting. The turbine work runs into delays, which then extend the outage, allowing the pump/valve mechanical contractor time for additional work. This contractor, in turn, discusses with the GC the need for certain valves to be replaced, instead of just repaired or repacked, especially since now there is time to order, receive, and install them. The mechanical contractor gets the OK and proceeds.

At the end of the job, the mechanical contractor submits an invoice to the GC for the extra cost of procuring and changing out the valves, and the GC, in turn, bills the owner. The owner asks for documentation that (1) proves the necessity of replacing the valves and (2) shows acceptance of agreeing to have the work performed. A good record-keeping system would enable the GC to access all correspondence necessary to satisfy the owner by keying into the GC's correspondence database a word or phrase such as "valve repair," which would immediately point to all documentation involved with this aspect of the work. The same goes for the mechanical contractor who pointed out the need for the work in the first place. There are many electronic systems available today that are designed specifically for this type of record retrieval, and the excuse for not being able to find the proper documentation is no longer acceptable.

Records such as the following should be kept for several years and be readily accessible (more on these in chapter 7):

- As-sold estimate
- All schedule revisions
- Daily progress report
- Transmittal letters
- Document status logs

- Clarification memos and RFIs (requests for information)
- Most correspondence, including e-mail
- Minutes of meetings
- Weekly and monthly status reports
- Photographs

It is critical to remember that all written and retained records are *discoverable* and not confidential. This means that once it has been written and recorded, any party to a future claim will have legal rights to see and use it. Note that individual (personal) diaries may also be admissible in court, so the best policy is to avoid using them. Sometimes, they contain comments of a personal nature, not very professionally written, that could result in some embarrassing statements being made public.

Finally, as part of the claims avoidance process, the prudent site management team should be aware of the Deadly Dozen, 12 causes of troubled projects that frequently lead to claims:[2]

1. **Delayed completion**

   When a construction project is completed late, the owner, contractor, and even the designer may experience financial damages. Unraveling delayed project completion is difficult and requires careful comparison between the originally planned schedule and the as-built version. Being able to plot some of the variables, such as personnel, percent complete, and the like, on the same time scale as the planned and as-built schedules will be helpful in seeing the cause of the delay. Therefore, these historic data should always be retained.

2. **Accelerated schedule**

   Acceleration of a project often comes about because of earlier delays that are being overcome. An accelerated schedule may result in increased cost to the party performing the work. More resources may be assigned, additional shifts may be implemented, and overtime may be required. Usually, the incremental cost for accelerating must be established, and the acceleration approved before extra payment will be made. This requires that cost records be available for verification.

3. Starting or ending date change

For most power plant construction work, especially outage-re-
lated work, starting and ending dates are carefully selected due
to seasonal power needs and climatic variations. Therefore,
the starting and ending dates of the project, as well as certain
activities within the project, should be carefully spelled out.
When changes are made in the time frames for the work—like
shifting the work into the hard winter or the very rainy season—
additional costs could be incurred. These incremental costs must
be established before extra payment is made. This requires that
records be available for verification.

4. Work sequence changes

The sequence of the work can be mandated by the contract
documents, or it can be implied by the nature of the work.
If there will be restrictions on the sequence of the work, the
party preparing the contract documents must specify this. Once
established, changes in the work sequence can be expensive
and time-consuming. However, good documentation must be
available to show that there was a required sequence, and if it was
unilaterally changed, that there were incremental costs.

5. Excessive management

Every project has some type of management structure. An
organizational chart should be prepared and distributed showing
the names of the companies and the individuals who are the
points of contact. Never skip these lines of communications,
and more importantly, *never* give directions to your contractor's subcontractor. Also, never direct any parties on how to
accomplish the tasks required by the contract, only state that
the task should be accomplished. Doing any of these can lead to
claims of increased costs due to excessive interference.

6. Lack of management by the owner or GC

The owner and/or GC has an obligation to coordinate the work
of the contractors and subcontractors, and this obligation must
be fulfilled. Lack of decisions in a timely manner will lead to
confusion, interference, and ultimately increased costs for which
claims can be made.

7. Quantity variations

When contractors bid and contract to do work on a unit price basis, they have to make some assumptions as to the quantity of work they expect to have available so they can properly charge for fixed costs like overhead and equipment. When there is a significant change from the estimated quantity of the work, there must be, by necessity, a method to change the unit price. In the case of unit price contracts, it is important to remain aware of the total quantities so proper adjustments can be made.

8. Quality of work

One of the most difficult things to define in a contract is quality. Quality is often determined by comparing the work on one project to similar work on another. In order to minimize disputes, the contract should invoke established codes wherever practical. If possible, sketch or construct a physical sample of the work desired and advise everyone that this is the quality required.

9. Access restrictions

Any access restrictions to the work area must be clearly spelled out in the contract documents. In many cases, especially when working in an operating plant, the bidders should be required to visit the area during operations to familiarize themselves with the restrictions so that there will not be any surprises during the performance of the work. If conditions change, there may be a claim for changed work sequence.

10. Failure of project to perform

A contract should be for *performance* or for *put in place*, not both. If it has some of both, confusion will result, and the parties will end up with claims against each other. As an example, if the electrical contractor is required to wire the drives of some equipment, say, large feedwater pump motors, in accordance with industry standards, then this contractor cannot also be told what size cable to use. However, if this contractor is told only what size cable to use, and this cable is too small, resulting in a fire, this contractor cannot be held accountable for the damage.

11. Additional costs because of the actions of others

Similar to item 4, a contractor can be impacted by the actions or inactions of others. This may be as simple as one contractor excavating across the access road to lay underground pipe, which then impedes the access to the work for everyone else. But it may also be as serious as the owner's employees blocking contractor access to the site as part of a labor action. Incidences such as these may cause contractors to incur additional costs.

12. Ambiguous contract documents

There are very few contracts that do not contain some ambiguity. Words like "timely," "prompt," and "workmanlike" should be redefined early in the project. Since one of the legal principles of construction contracting is that an ambiguous clause will be interpreted against the person who prepared the contract, it behooves this party to review the contract wording and clarify all ambiguities.

In summary, with the availability of all of the foregoing records, data, and other information, the on-site parties should be able to formulate claims, defend from claims, and ultimately resolve claim issues amicably. Ideally, the claims can be settled on-site between the parties. But when they cannot, another process must be started because the people at the site do not have the time to process contentious claims and still get their jobs done. So claims are often sent on to others for processing.

Once a claim leaves the site, it should be treated like a project. It needs to be worked by a team with a leader. Like any project, it should be defined, scheduled, budgeted, and managed. It should also go through three phases. Phase 1 is an evaluation of the claim, which is accomplished by reviewing selected project documentation and identifying the issues in dispute. A risk analysis follows, resulting in a decision whether to go forward or not, based on the potential of success versus the cost of pursuing it. Phase 2 is a full-fledged analysis of the claim with solid positions supported by credible documentation and knowledgeable personnel. Phase 3 is the resolution procedure itself, something that can take many forms.

Depending on the perception each party has of the strength of the other's position, claims may be resolved through party-to-party negotiations. When that does not work, arbitration can be used. Often, mediation is selected as a cost-effective way to resolve claims. And then, if all else fails, there is still the court system—litigation. However, all of these efforts are costly, and they often create ill will. The preferred way to handle this is to avoid claims from the outset. Understand the contract, clear up its ambiguities, make timely notification of issues, then discuss them—and always document, document, document.

A final word needs to be said about contentious claims and the need to avoid them whenever possible. Just because a contract is worded in a specific way, it may not be enforced that way if allowed to go to litigation. Here are a couple of examples:

- Inordinate delays. If the contractor is forced to perform under more difficult and costly conditions due to delays outside of the contractor's control, the courts may allow recovery for the excess costs, even though the contract clearly stated that the contractor would not be allowed recovery for *any* cost overruns. For example, a power plant located in the far north did not shut down during the fall season, as the contract said it would; rather, it shifted its outage to the winter, with the accompanying severe weather and holidays. Many, although not all, courts will use a sense of fairness when adjudicating a claim for recovery of the extra costs incurred due to this *inordinate* type of delay. One never knows who the presiding judge will be!

- Unit rates. Many contracts have a provision for using fixed unit rates to adjust the price in the event of additions or deletions. However, if the change is of such a nature that the unit rate is no longer representative of the work required, the court may not uphold the contractual rate. For example, if the unit prices for the erection of a ton of steel are fixed and a change in the design greatly increases the number of pieces of steel per ton, thereby increasing the cost of installation of this ton of steel, then the unit prices may not be upheld. Of course, this could also work in reverse.

- Cardinal change. Although the contract may clearly state that the contractor is required to install all items or equipment necessary for the completion of the work, the courts may impose limits on this. For example, let's say the insulation installation contractor was told that there would be a certain quantity of 3-inch insulating material to use for preparing an estimate to insulate a boiler, and let's assume that the contract had a clause similar to the one previously noted. However, once the job got underway, the material supplied was actually two layers of 2-inch insulating material, which required a *substantial* increase in labor to install. Regardless of the wording of the contract, the courts may find in favor of the contractor since there was a drastic or substantial increase in the change of the work, a concept known as a *cardinal change*.

Insurance Management

If contract language and claims are two of the legs of the risk mitigation stool, the umbrella of insurance is the third leg. The use of insurance as a tool to recover from unexpected costs due to unforeseen events is a common risk-mitigation approach. Just like managing the power plant construction process to avoid costs from claims, protection from unexpected and accidental risks, including black swan events, must also be provided for and managed. This is usually done in concert with the insurance industry. Construction work, by its very nature, is a risky business. Aside from the personnel safety issues, which will be covered in chapter 9, there are a host of other risky and potentially costly issues. The first, and most important risk is the stability of the partners of the project, both owners and contractors. The next is a litany of accidental risks, such as physical damages, third-party issues, automobiles, and others. Although most of the mitigation effort for these risks is done long before the site mobilizes, it is important that the site personnel are familiar about the protections in place, their scope and limitations, and also the actions that the site personnel must take to avail themselves of the coverage provided. The next few sections will provide an overview of the typical risks that are usually insured and what must be done at site to keep this protection in place.

Surety bonding

Gambling on a contractor or subcontractor, whose level of commitment is uncertain or who could become bankrupt during the job can be an economically devastating decision. With most new power plant construction projects, as well as all outage work, being under pressure to complete with the lowest possible dollar cost to the project, contracts are often awarded to the lowest priced bidders. Unfortunately, the lowest price does not always result in the lowest cost. So how is the owner or GC supposed to be sure the lowest price stays that way? The most common process used today is to use a surety bond.

A *surety bond* is a written agreement where one party, the surety, obligates itself to a second party, the obligee for the default of a third party, the principal. In the case of power plant construction work, a surety bond provides financial security and construction assurance to the owners that the contractors will perform the work and pay their subcontractors, craftsmen, and material suppliers. It is a risk-transfer mechanism where the surety company assures the owner (obligee) that the contractor (principal) will perform in accordance with the contract documents. It offers assurance that the contractor is capable of completing the contract on time, within budget, and according to specifications.

There are alternative forms of financial security, such as self-insurance and letters of credit, but these are not as comprehensive as a surety bond. Almost all publicly held utilities are mandated by law to use surety bonds, and many private owners also require them. With surety bonds in place, the risk of project completion is shifted from the owner to the surety company, protecting both the owner company and its shareholders from the enormous cost of contractor failure. Subcontractors are also often required to obtain surety bonds to help the prime contractor manage risk.

Most owners and GCs will require three basic bonds. The first is to ensure that the contractor will stand behind the submitted bid in the event of an award. This is called a *bid bond*. The second, called a *performance bond*, is to ensure that the project will be completed as provided in the contract. This is the heart of surety bonding. A third bond is usually required to ensure that the contractor pays all of the hired personnel, suppliers, and subcontractors and that they will not place a lien against the property. This is called a *payment bond*.

The first of these, the bid bond, is usually not of concern to the job-site personnel. By the time they arrive on-site, the contractor has usually been selected and the contract signed that then releases that bond. However, the second and third bonds, performance and payment, do require cognizance by the site staff. The performance bond, for example, is usually a function of the value of the contract. As the job progresses and changes are authorized, the value of the contract may also change, requiring notification to the surety company so the bonding value can be adjusted to maintain adequate protection. Often, the site personnel are responsible for notifying the individuals responsible for maintaining this coverage.

As opposed to insurance policies, which are written with the expectation of a number of losses, surety bonds are written with the expectation of only a few. Because of this expectation of only a few losses, surety companies will perform a rigorous examination of the contractor before issuing coverage. They will investigate to be sure that the contractor has the following:

- Good references and reputation
- The ability to meet current and future obligations
- Experience matching the contract requirements
- The necessary equipment to do the work or the ability to obtain it
- The financial strength to support the desired work program
- An excellent credit history
- An established bank relationship and line of credit

Because the intent of the surety bond is to protect the owner and/or prime contractor from the potentially devastating expense of contractor and subcontractor failure, if any of these criteria change, the surety company must be made aware of these changes. It is incumbent upon the bonded contractor to advise the appropriate individuals in these cases. For example, if the project requires some heavy construction equipment central to the performance of the work, let's say an 800-ton crane, and suddenly this equipment is no longer available, the contractor must work with the contractor's surety to come up with a solution that allows the job to proceed. Surety companies can prevent default on contracts by offering technical, financial, or management assistance. But they must be offered the opportunity to participate at the beginning, when the issues first surface.

Sometimes, even the best efforts of the contractor and the surety are not enough to prevent default. As mentioned before, construction is a risky business. In the event of contractor failure, the owner or GC must formally declare the contractor in default. When this happens, the surety will conduct an investigation. Once it has been determined that the default is real, the surety's options, which are usually spelled out in the bond, are invoked. These options may include the right to rebid the job for completion. They may include the surety bringing in a contractor of its own choosing to complete the job. Another possibility is for the surety to provide financial and/or technical support to the defaulting contractor. And if all else fails, the surety can pay the owner the penal sum of the bond, leaving the owner with the task of completing the job using this money (which may be insufficient).

As is very obvious, having to engage the surety in the salvation of a project is not desirable. The preferred way is to use viable contractors and solid contract language to set up the job and then to work with the contractor to reduce the possibility of major problems.

The following is a classic case of not working together, and then having a *lose–lose* situation. The GC hires a subcontractor for a major portion of the work. The subcontractor arranges for performance and payment bonds, as required by contract. As the job progresses, issues arise that become contentious and lead to potential schedule delays. In an effort to pressure the subcontractor to get back on schedule, the GC starts withholding progress payments, in the guise of invoking liquidated damages (LDs). This now puts a strain on the subcontractor's cash flow. Suddenly, the subcontractor has trouble paying the labor and suppliers. The labor, in turn, no longer performs at their peak, and suppliers stop extending credit. What happens next?

With reduced productivity from the craftsmen, and with fewer supplies with which to work, the subcontractor gets further and further behind. The project is now definitely behind schedule. The owner becomes concerned and pressures the GC by threatening to invoke the contractual LDs, which are much higher than those the subcontractor has with him. Finally, the surety is called in, the subcontractor either declares bankruptcy and abandons the site or is removed from the job site, and the GC and surety have to work out a way to get the job completed.

The results are as follows:

- A delayed job that prevents the owner from generating power and receiving revenues
- A GC who has been forced to pay LDs to the owner
- A subcontractor now either in bankruptcy or at least with a tarnished credit rating and reputation, affecting future business prospects
- A surety having to pay substantial sums to back up a bonded promise to complete the job, which will result in higher premiums the next time, leading to increased costs for everyone in the future

Why did all of this go so wrong? Simply because the GC started withholding payments from the subcontractor, instead of working with the sub and the surety to get the job back on schedule. With personnel on-site who understand the bonding process and who are aware of the future consequences of not notifying the surety as soon as issues arise, situations like this can be minimized. However, since most site personnel do not understand the subject of surety bonding, some basic training is often necessary.

Finally, a few words on payment bonds. This is the third surety bond usually provided on most sizable construction projects, whether it is a new plant or a major outage. The labor and the material and equipment suppliers that a contractor uses expect to be paid for their services and products. Usually, they have provided these services and products long before being actually paid for them. Especially with third-party vendors, they may have extended significant amounts of credit, for which they anticipate being repaid in a timely fashion. If for some reason, the contractor leaves the job without paying these people, they usually have the right under most legal systems to place a lien against the installed work, essentially preventing the owner from putting the plant into operation.

To prevent this situation, owners frequently require a payment bond in addition to the performance bond. This payment bond, usually backed by the same surety company that provides the bid and performance bonds, can then be invoked by the owner to pay off the contractor's creditors, resulting in the liens being removed.

Smaller projects, such as outage turn-arounds, may not have performance bonds. In these cases, it is common for the contract terms of payment to require a withholding of 10% or 15% of the contract value, which will not be paid until the contractor provides a "release of lien." This release is usually a legal document affirming that all suppliers have been paid, that the plant is free from liens and that the contractor will defend and indemnify the owner from any future claims or liens related to the work (fig. 6–5). Alternately, some owners will require a waiver of lien separately from the contractor and each of the contractor's suppliers.

LIEN AND CLAIM RELEASE FOR ORDER NO._____

Conditional upon payment to Seller, the sum of _____ dollars ($_____) by _____, hereinafter "Purchaser", _____, hereinafter "Seller", does for itself, its successors and assignees, hereby release and discharge Purchaser, its officers, agents and customers from any and all claims, demands and liabilities whatsoever arising under or by virtue of the referenced Order.

Seller covenants and warrants that the premises on or for which the Work was performed; services rendered and materials furnished are free from all liens and claims chargeable to the premises by reason of Work performed, services rendered and materials furnished by Seller and by any subcontractor, supplier, employee and agents working for or under Seller. Seller agrees to indemnify, protect and save harmless Purchaser and its customer from any claims or demands for Work performed, services rendered and materials furnished by Seller under the referenced Purchase Order and to defend all actions arising out of said transaction and Seller shall pay any costs and expenses including reasonable counsel or attorney fees incurred by Purchaser or its customer in defense or settlement of any such claims and demands.

_____(Seller)

Attest:

By:

Title:

State of

County of

On this _____ day of _____, 20_____ before me, the subscriber, personally appeared _____ to me personally known and known to me to be the same person who executed the within instrument, and duly acknowledged that he executed the same.

_____

Notary Public

Fig. 6–5. A typical lien and claims release form (Courtesy of Construction Business Associates LLC)

Other insurance

Bonding the contractor to ensure performance is not the only insurance needed to manage the risks of the site work. There are many other risks that, if not properly managed, also could cause a job to deteriorate quickly. The following are some of the more common forms of insurance for power plant construction projects, but they are not the only ones that might be encountered:

- Marine transit. This is insurance protection in the event that materials being transported to the job site are damaged or lost at sea.

- Builders' all risk (BAR). This provides protection for the owner, contractor, and subcontractors in the event of loss or damage to materials, supplies, and equipment, as well as work put in place, from all types of causes, usually also including hurricanes, floods, and earthquakes. However, there are still some exceptions in every policy, such as theft.

- Third party and general liability. This protects the builder from claims of injury to a third party as a result of the work. Without this insurance, the owner could be held liable.

- Workers' compensation. This covers anyone employed by the contractor in the event of injury during the course of work for the contractor. Without this insurance, the owner could be held liable.

- Automobile liability. This insurance is for protection from liability in the event of damage to automobiles or to third-party property, or in the event of injury to personnel as a result of an accident involving the contractor's vehicles.

Although obtaining the above types of insurance is a prudent risk management tool, there are loopholes that must be clearly understood. For example, a BAR policy usually provides for reimbursement of repeat or extra work in the event of a loss to the work. However, if the loss to the work was due to an inadequate design, the insurer usually will not pay. This situation could arise in the event of the collapse of a structure, say, a coal silo supported on structural steel, which was determined to be underdesigned. (Sometimes these damages may be covered by a professional errors and omissions policy, but that is beyond the scope of this book).

In the old days of power plant work, each contractor on-site usually provided all of these forms of insurance. This practice created duplicate coverage in many instances; because of today's focus on the bottom line, this has changed. Although the insurance protections are still provided, there has been a shift to consolidation. The owner will now sometimes provide some of this insurance coverage for itself and most of the contractors on-site, often referred to as a *wrap-up* policy or owner-controlled insurance program (OCIP). There are significant savings in premium payments by the owner purchasing one policy and having all of the affected contractors named as additional insured.

While saving the project the cost of multiple premiums, OCIPs also must be clearly understood by the contractors. Generally, there are deductibles that may be larger than some contractors care to encounter, so they may still want to obtain insurance coverage up to the level that the OCIP begins. Also, not every OCIP includes all of the above listed types of insurance. Classic omissions are workers' compensation and automobile liability. If these are not included in the OCIP, the contractors must procure them separately. (See chapter 9 for more on OCIP.)

For these forms of insurance to be effective, the site management—whether the owner's staff, the contractor personnel, or the subcontractors—must be aware of the coverage, and they must know the requirements of notification, implementation, and documentation. Improper or delayed notification may partially or wholly invalidate coverage. The insurer usually reserves the right to mitigate the insured damage in ways that are in the insurer's best interest. If the insurer is not given timely notification of the event, it may no longer be able to rectify the situation in a manner that is cost-effective to it.

To ensure that the insurance coverage is as required and is actually in effect, most owners will not allow any contractors to start work on-site until proof has been provided of coverage and proof that the coverage will remain in force until the job is complete. This type of proof is usually in the form of a certificate of insurance that outlines the insurance and limits provided, names the insurer providing the coverage, and also specifically names all of the additional parties that are covered (the additional insured) (fig. 6–6).

ACORD™ CERTIFICATE OF LIABILITY INSURANCE

DATE (MM/DD/YYYY)

PRODUCER	THIS CERTIFICATE IS ISSUED AS A MATTER OF INFORMATION ONLY AND CONFERS NO RIGHTS UPON THE CERTIFICATE HOLDER. THIS CERTIFICATE DOES NOT AMEND, EXTEND OR ALTER THE COVERAGE AFFORDED BY THE POLICIES BELOW.
	INSURERS AFFORDING COVERAGE NAIC #
INSURED	INSURER A:
	INSURER B:
	INSURER C:
	INSURER D:
	INSURER E:

COVERAGES

THE POLICIES OF INSURANCE LISTED BELOW HAVE BEEN ISSUED TO THE INSURED NAMED ABOVE FOR THE POLICY PERIOD INDICATED. NOTWITHSTANDING ANY REQUIREMENT, TERM OR CONDITION OF ANY CONTRACT OR OTHER DOCUMENT WITH RESPECT TO WHICH THIS CERTIFICATE MAY BE ISSUED OR MAY PERTAIN, THE INSURANCE AFFORDED BY THE POLICIES DESCRIBED HEREIN IS SUBJECT TO ALL THE TERMS, EXCLUSIONS AND CONDITIONS OF SUCH POLICIES. AGGREGATE LIMITS SHOWN MAY HAVE BEEN REDUCED BY PAID CLAIMS.

INSR LTR	ADD'L INSRD	TYPE OF INSURANCE	POLICY NUMBER	POLICY EFFECTIVE DATE (MM/DD/YY)	POLICY EXPIRATION DATE (MM/DD/YY)	LIMITS	
		GENERAL LIABILITY				EACH OCCURRENCE	$
		☐ COMMERCIAL GENERAL LIABILITY				DAMAGE TO RENTED PREMISES (Ea occurence)	$
		☐ CLAIMS MADE ☐ OCCUR				MED EXP (Any one person)	$
						PERSONAL & ADV INJURY	$
						GENERAL AGGREGATE	$
		GEN'L AGGREGATE LIMIT APPLIES PER: ☐ POLICY ☐ PRO-JECT ☐ LOC				PRODUCTS - COMP/OP AGG	$
		AUTOMOBILE LIABILITY				COMBINED SINGLE LIMIT (Ea accident)	$
		☐ ANY AUTO					
		☐ ALL OWNED AUTOS				BODILY INJURY (Per person)	$
		☐ SCHEDULED AUTOS					
		☐ HIRED AUTOS				BODILY INJURY (Per accident)	$
		☐ NON-OWNED AUTOS					
						PROPERTY DAMAGE (Per accident)	$
		GARAGE LIABILITY				AUTO ONLY - EA ACCIDENT	$
		☐ ANY AUTO				OTHER THAN EA ACC	$
						AUTO ONLY: AGG	$
		EXCESS/UMBRELLA LIABILITY				EACH OCCURRENCE	$
		☐ OCCUR ☐ CLAIMS MADE				AGGREGATE	$
							$
		☐ DEDUCTIBLE					$
		☐ RETENTION $					$
		WORKERS COMPENSATION AND EMPLOYERS' LIABILITY				☐ WC STATU-TORY LIMITS ☐ OTH-ER	
		ANY PROPRIETOR/PARTNER/EXECUTIVE OFFICER/MEMBER EXCLUDED?				E.L. EACH ACCIDENT	$
		If yes, describe under SPECIAL PROVISIONS below				E.L. DISEASE - EA EMPLOYEE	$
						E.L. DISEASE - POLICY LIMIT	$
		OTHER					

DESCRIPTION OF OPERATIONS / LOCATIONS / VEHICLES / EXCLUSIONS ADDED BY ENDORSEMENT / SPECIAL PROVISIONS

CERTIFICATE HOLDER	CANCELLATION
	SHOULD ANY OF THE ABOVE DESCRIBED POLICIES BE CANCELLED BEFORE THE EXPIRATION DATE THEREOF, THE ISSUING INSURER WILL ENDEAVOR TO MAIL _____ DAYS WRITTEN NOTICE TO THE CERTIFICATE HOLDER NAMED TO THE LEFT, BUT FAILURE TO DO SO SHALL IMPOSE NO OBLIGATION OR LIABILITY OF ANY KIND UPON THE INSURER, ITS AGENTS OR REPRESENTATIVES.
	AUTHORIZED REPRESENTATIVE

ACORD 25 (2001/08)   © ACORD CORPORATION 1988

Fig. 6-6. Certificate of liability insurance (Courtesy of Acord)

Summary

Managing the risks of a power plant construction process is a very important part of managing the overall project. As some of the examples in this chapter have shown, issues can arise in spite of the best efforts of the site management teams. Issues also can arise due to the direct actions of the site management teams. Things change during the construction process.

Some of these changes result from one party's actions against the other. Some of the changes are the results of third parties and some are due to causes outside of the control of any party. But whichever form change takes, or from wherever it comes, it must be addressed and managed.

The tools to manage change are the claims process and the insurance process. The claims process is dependent on the contractual clauses agreed between the parties, usually long before the site team has even been selected. However, the site team must become familiar with these clauses so that (1) actions can be taken in time to avoid claims, and (2) when claims issues do arise, they are handled appropriately.

As the project progresses, communications must be made an integral part of the process. When even the hint of an abnormal condition arises, the parties to the contract should review the issue and look for ways to resolve it. Allowing issues to move forward without proactive intervention often results in costly claims and disgruntled people.

The insurance process is also dependent on agreements usually made prior to the site team selection. Differing from the claims process, however, the insurance process looks to third parties for resolution and protection.

Since most construction contracts are estimated, bid, and awarded based on known and predictable events, the costs of unexpected issues are not included. However, unexpected events do occur, and since they do, they must be managed in such a way as to minimize their impact on the project. This is done through the use of insurance companies bonding contractors for performance and insurance companies providing financial protection in the event of unexpected incidences or unforeseen perils—sometimes called black swan events.

Although the site management does not need to have insurance experts on its staff, it must have staff familiar with insurance concepts. They must know what to do to protect the interests of the insured as well as the insurer. They must know when and whom to notify of events that might trigger a claim. They must understand the interrelationships among the owner, contractors, and subcontractors.

In summary, managing the day-to-day construction operations of a power plant project, whether it is a new plant or a retrofit, is complex in and of itself. But add to that the realities that nothing ever remains the same and that things are not always as they seem, and suddenly a host of unplanned, unexpected issues arise that must be managed to prevent a collapse of the project.

These risks require intelligent planning, often long before the site even mobilizes. They require smart site managers who understand the potential for damage on-site. These risks require careful managing, in accordance with defined parameters, and they require teamwork. The unexpected can be tamed!

References

1 McCue, Robert C. "Complexity as Culprit." *Engineering News-Record*, June 25, 2012, p. 88.

2 Wilson, R. L. *Claims Avoidance: Team Work for Positive Results*. Glen Head, NY: Wilson Management Associates. November 2000.

Setting Up the Job Site 7

There comes a point when it is time to put to the test all of the thoughts and dreams, the hopes and fears, and the hard work and due diligence that went into preparing for the actual project execution. There comes a point when it is time to put the shoe leather to the road and get this construction job underway.

Once the contract negotiations are over and all the parties have come to terms, both on technical issues and commercial conditions, the next step is to write down an action plan, called a project execution plan (PEP) or a project or construction management plan. Whatever you call it, the operative word is PLAN. It must be in writing, and it must cover certain salient points.

Often, this is when the site manager is appointed. Therefore, the site manager may not know much about what transpired during the earlier phases of the project. Many of the earlier steps that led up to this point in the project may have been taken without the site manager's involvement. Steps such as the lessons learned sessions during the preplanning phase (chapter 1) probably were conducted long before the site manager was hired. The same goes for determining the project delivery structure—is this project a joint venture? Is it direct hire or subcontract?

What is the budget, and how was it derived (and by whose standards)? Are there contingency plans, in both dollars and time, in the event that things do not go as planned? What about the schedule—is it realistic today? How about payment terms? Are there penalties? Are there bonuses?

Where is the labor intended to be sourced from? What about the supervision? And are those all-important tools, equipment, and consumables available as needed and when needed? And finally, what is in those terms and conditions (T&Cs) that might jump up and catch us off guard?

It is a lot to think about for a site manager when first assigned to a project. A site manager who was part of the planning team from the beginning would know where the traps lie. This site manager would know where the risks might occur and where there is some cushion. But even then, a lot may have changed since those initial days of preplanning and planning, and many items may have to be revisited and/or revalidated.

The Project Execution Plan

There are probably as many different ways to develop a PEP as there are people available to develop one. As a quick digression, the person responsible for the total project, from inception through engineering, from procurement to material and equipment delivery, and from construction through commissioning and start-up, will have a PEP that is all-encompassing. This person will most likely look to each major project executor—engineering, procurement, construction, and commissioning/start-up—to have their own PEPs to be integrated with the overall project execution plan and philosophy. For purposes of this book, the discussions here will be about the PEP for the construction phase only.

During the development of the PEP for the construction phase, attention to details will make the management of the site much easier. In fact, the whole purpose of the job-site setup is to facilitate the management process. Time should be spent thinking about all of the steps in the process of managing the site activities, before the actual labor efforts begin. Thought needs to be given to the administrative requirements, the actual site project management tools that will be used, the purchasing process, how quality will be measured and controlled, and how safety will be managed. This should be captured in the PEP for the construction work.

Site Administration

With the details of the contract final, the process of setting up the actual site organization can start, and this means organizing the site administration. To ensure an orderly flow of information and an effective process of project controls, it is extremely important that the administrative processes be properly designed. The basic administrative needs of most construction projects include administering the supervision, the craftsmen, the payroll, the field office, the materials, and the tools, facilities, and equipment. And of course, there is the documentation, or record keeping, that must also be established. But first, the basic needs:

- The task of administering the supervision, although not major, should be well planned. It is important to realize that the supervisory personnel set the mood of the project; their enthusiasm, or lack thereof, is directly imparted to the craftsmen whose attitude, in turn, reflects directly on the work they perform. If the supervision is unhappy or not supportive of management, the project will be an uphill battle.

  Therefore, job assignment benefits, such as per diem payments, insurance provisions, automobile policy, housing allowances, single/married status, meals, home leave, and so forth, need careful review. The benefits must be attuned to the project scope, size, location, and duration, and they also must be reflective of the individuals and their needs. In other words, job benefits need to be meaningful for the individual, but they cannot discriminate against others.

- Administering the craftsmen requires consistent coordination between the field supervisors and the office administrative staff. The craftsmen's function at the job site is to perform their tasks as expeditiously as they can. If they have their minds on other issues such as unfair task assignments, pending layoffs, safety matters, or even poor payroll procedures, their performance will be affected. Therefore, the field supervisors must make job assignments in a consistent and fair manner, they must treat the craftsmen with respect, and they must work within the job-site rules and union regulations, where applicable. At the same time, the office administrative personnel must ensure that consistent procedures are followed as well. They must be followed from the start of each shift with the "brassing in," or, as it is often referred to these days, "badging in," using electronic swipe cards, through to the calculation of each craftsman's paycheck at the end of each pay period.

  If the labor force is unionized, additional interfaces are required to ensure compliance with labor–management agreements. The job-site rules must be clear before the start of work assignments and the availability of skilled craftsmen must be ascertained. An open-door approach between the union and project management will facilitate the resolution of misunderstandings throughout the job. Regular, tripartite meetings with the owner, the contractor, and the union representatives attending will go a long way toward defusing volatile issues and setting the tone for the tradespeople and supervisors to work in harmony.

- Payroll administration can be a complex process. In addition to internal company requirements, there are local, state, and federal rules. The simple task of applying for a job by a candidate requires confirmation that the applicant is not only capable of performing the work but also legally allowed to do so. It is important that the weekly payroll is calculated accurately, that the withholdings are in accordance with all legal (and union) requirements, and that supporting documentation exists. Time sheets must reflect the actual hours worked, differentiating between overtime and straight time; the pay must be calculated in accordance with the time sheets; the checks or direct deposits must be issued regularly; and the payments to the local, state, and federal governments must be made timely and accurately. The same holds true for payments to the unions. In addition, year-end reporting requires that accurate records be maintained even though the job may have been completed many months before. The banking process needs to be seamless. A timely notification of anticipated payrolls needs to be communicated to the dispensing authority so that funds are available in time for workers to cash their paychecks or for electronic transfers to be honored.

    Finally, the basis of the project control process is usually a function of the man-hours expended compared to the man-hours still to be spent. The payroll process is generally the easiest place to record and categorize the man-hours expended. This may be done by the use of sophisticated computer programs; on smaller projects it may be done using simple spreadsheets. Whatever the method, if it is not used correctly and consistently, the status report of the job will not be correct and the projection to completion will be meaningless. (More on this is in chapter 10.)

- The field office is the center of the site administration. By properly setting up this office, including all of the record-keeping requirements, the task of the supervision and management will be more efficient and pleasant. A good summary of the salient points of the contract, sometimes referred to as a project abstract, is a useful guideline for setting up the various files. Correspondence files can be arranged to readily support requirements such as reporting, delay notifications, nonconformances, extra work authorizations, billings, and insurance claims.

  Equally important as setting up the field office at the beginning of the job is closing it down at the end. The contract may stipulate what records should be turned over to the client, the owner may have specific requirements regarding disposal of remaining materials and specialized tools, and internal company procedures may have certain regulations, especially with respect to record keeping, that are necessary to maintain compliance with legal, code, and other quality control requirements. Additionally, there may be special regulations affecting the retention of safety records.

- Material control is one of the more critical aspects of the job since without materials to install, the job comes to a halt. Therefore, it is important to have a good material control program; generally, a separate material control person will be required. The material control program should enable the field supervision to determine when specific materials will arrive at site, and that they have actually arrived at site, in what condition, and where on-site they are stored. In today's world of electronics, these tasks can be greatly simplified by using a bar code or similar system. (More on this is in chapter 13.)

- Administration of the job-site tools, facilities, and equipment is an area where often money is spent needlessly. Since 15% to 20% of the job budget is for these items, proper attention to setting up the controls will save aggravation and money when the job starts closing down. The job-site management should procure tooling and equipment in keeping with their availability and the skills of the labor and supervision that will use them. Usually, there is more than one way to accomplish many of the tasks in the building of a power plant, so it is important to investigate the cost of trading man-hours for the use of sophisticated tools and equipment. Also, on large overseas projects, shipping time and costs, as well as demurrage and customs procedures, may impact the decision more heavily than just the savings in labor efficiency.

Once the list of tools, facilities, and equipment has been finalized, a look at the accounting rules is prudent. Depending on the value and useful life of the tools and equipment, some of them may be depreciated over several years instead of being charged 100% to the project, thereby sparing the job the total cost of these items; this then enhances the profitability of the job. The same holds true for a decision to rent versus purchase. As the job winds down, a concerted effort should be made to inventory all remaining tools, facilities, and equipment. Items not required any longer should be disposed of, and those still being used should be placed on a "watch list" so they can be removed as soon as they are no longer needed. Additionally, many of the more expensive pieces of equipment, like large trucks and cranes, can be replaced with less costly equipment, or even with personnel. Often, it is very cost-effective to assign a dedicated individual to expedite the removal of the tools, facilities, and equipment as the job starts winding down.

A final note regarding tools and equipment: Sometimes it is desirable to loan or rent these items to third parties. Good business practices suggest that if this is done, proper legal and insurance protections are put in place, such as a hold harmless statement.

Complementing the above, record-keeping requirements of the project must be addressed. A typical job may require a set of documents as follows:

- Daily progress reports. There are many formats that can be used, but the format adopted must be able to capture the critical elements of the specific job. For example, it may not be important to document the daily weather when the work is all indoors, but if you are erecting an outdoor unit in the winter in snow country, weather documentation may be the difference between the imposition of a $300,000-per-day delay penalty versus an extension of time!

- Transmittal letters. Everything sent to other parties, whether it be drawings, change orders, pay requisitions, or other documents, should be sent with a transmittal form or letter, hard copy or electronic. This transmittal form should include the name of the recipient, the name of the sender, and the date it is being sent. If a response is required, it should be noted on the form. Often it is also desirable to have the recipient sign an acknowledgment of receipt, either manually or electronically.

- Document status logs. At the beginning of the job, a system of documenting the flow of the various documents of the project should be implemented. This may be a manual process for smaller projects, but on complex jobs, it should be a database system with the ability to provide information on what documents have not yet been processed, report the status of those that are in process, and sort on specific words or topics for historical information.

- Clarification memos or RFIs (requests for information). Throughout the course of a project, clarifications are often needed from the owner or designer. These requests should always be in writing and dated, and the responses should be logged.

- Correspondence. Keep correspondence simple, and use only one letter, memo, or e-mail per subject. Then be sure it is logged and filed accordingly.

- Minutes of meetings. There should be an agenda for every meeting, and minutes should be prepared shortly after the meeting is complete. These minutes should then be issued to all interested parties and filed according to topic.

- Status reports. Contractors should provide weekly and monthly status reports on every job. It is a good way to force the contractor to think through the events of the past and prepare for the next period in a logical, systematic manner. If the job is a fast-paced outage, progress reporting generally needs to be done on a daily basis. This usually works best when all parties involved in the project meet at the same time, in the same place, every day. The information from these sessions should then be applied immediately to the master project schedule and a determination made to see if the project requires any tweaking to keep it on schedule and within budget.

- Photographs. There is no better way to document what is happening on a construction site than with photographs or videos. By using continuous video and transmitting photos electronically, problems and misunderstandings can be demonstrated in ways that are much more efficient than letters or marked-up drawings.

Finally, if in the past there have been similar projects, a lessons learned session should be held with the participants of the old as well as the current project. Some of the ideas that may come from such a session are a need to identify employees of different contractors, or contracts, by color coding their hard hats. Trade-offs between subassembling components and installing individual parts can be discussed, possibly saving many hours of labor. Ideal use of heavy equipment such as costly cranes should be debated because frequently this is an area of cost overruns. Another idea for discussion is to review the feasibility of installing scaffold brackets, and perhaps the scaffolding onto large components, before raising these components, avoiding the need to do this once at elevation. Decisions made at this stage of the project will have a major impact on the financial outcome of the job.

In today's world of high-speed information technology, the flow of the information generated by the site administrative processes should be automated. The high cost of power plant construction today and the costs of not meeting the schedule and budget commitments that were made demand that management must have pertinent information to make decisions 24 hours a day, seven days a week. This can be accomplished readily using a multitude of tools, ranging from simple e-mail attachments to dedicated centralized servers and/or cloud storage, which can be accessed by many. The recipients are usually less concerned with the method of information transmittal than they are with the timeliness of receipt.

Managing the Site

Some construction sites are huge, others are not, but both are often terribly congested, fragmented, and in a state of total flux. The amount of time that can be saved by setting up a clear, streamlined site management process is tremendous. This goes all the way from where the workers park and where they eat to how they and their tools and equipment move around the site.

Traffic flow

The first item of business that should be addressed at the initial site-planning meeting between the owner and contractors is how the traffic will flow. With a site plan and the project schedule in hand, all parties should be prepared to discuss how they intend for their materials to move from unloading to laydown and from the laydown area to the site and on the site. The same should be done for major equipment. This accomplishes several things. One is to establish what type of road beds need to be available and when they need to be available. Another is to force a preliminary review of construction access needs. In other words, areas that need to be left open for crane or other large vehicle movement are highlighted—for example, leaving steel out, not erecting certain buildings yet, or delaying the excavation for pipe chases and installation of elevated cable trays.

Next, ingress and egress of the workers must be addressed. Where will they enter the job site, and how: by foot, bus, or car? How many workers will there be at the different stages of the work, and what will be the job-site obstructions then? What about working hours: Should there be staggered start and stop times? Two entrance gates for large jobs could be the solution to crowded shift changes. Where will the changing rooms be located, where will the sanitary facilities be, where will the workers eat, and very importantly, where will the first aid and ambulance service be stationed? Also, if elevators are used, how is the traffic regulated? Are there enough elevators, or are there workers standing around for 15-minute intervals, being paid to wait for the next empty elevator car? Again, diligent preplanning will help to minimize a lot of congestion and increase productivity.

For job sites with a high density of equipment movement and/or personnel, it sometimes pays to have one person dedicated to traffic control. With cumulative job-site payrolls of over $500,000 per day on large projects, saving one minute of confusion translates into $1,000, which is much more savings than the cost of the traffic controller.

Site housekeeping

There is no worse eyesore than a construction site where all kinds of buildings and structures are just partially erected, and there is a swirl of paper, cups, rags, and other debris blowing around on a windy day. It is emotionally depressing and physically unsafe. Add to that spent weld rod ends, broken pallets, and dunnage lying around. Then note the sloppy welding leads, the haphazardly strung electrical cords, and even workers' jackets, hard hats, gloves, and lunch bags strewn around, and what do you have? A place where no one looks forward to coming to work; a place that is ripe for an OSHA (Occupational Safety and Health Administration) violation and a place that is certainly not efficiently run.

As with traffic control, site housekeeping is an issue that affects everyone. There is seldom a project of any magnitude where the craftsmen who make the mess are the only ones who see the mess. Usually, many different workers, from many different contractors, use some of the same areas at the same time. Stepping over obstacles, avoiding dirty areas on the way to the workstation, or stopping to move cables and slings that are in the way create unnecessary delays for the person on the way to do a job. It is simply inefficient and potentially unsafe.

The most effective way to keep a site clean is for each employer to train workers in housekeeping and to emphasize cleanliness in safety meetings and during lunch breaks or shift changes. However, this is never enough. A separate housekeeping crew, with workers from each major contractor, is often used to make regularly scheduled rounds of the premises to police the area. Often, this is under the supervision of the safety officers on-site, but it could also be led by a foreman from one of the contractors, rotating on a weekly basis. The price is small when compared to the cost of the inefficiencies and the impact from a safety standpoint.

Site services

Almost no power plant construction site contractor is self-contained. Someone else is providing the power. Someone else is providing the construction water. Someone else is providing the dust control, and someone else is also providing the trash removal. Often, that someone else is the owner. On a very large project, it may be the general contractor. But usually the lower-tier contractors depend on someone else to provide these services. The issue that usually arises is the adequacy and reliability of having these services where they are needed, when they are needed.

For example, let's look at the temporary power supply. It is uncanny how a power plant site can lose power just when it is needed most—at the peak of production welding on the boiler when the hydrotest date is in jeopardy. Whether the job is to construct a new unit or to overhaul an old one, the calculations for determining the peak power requirements are not rocket science. A simple summation of the maximum power requirements from all of the contractors, superimposed on the construction schedule, will readily determine what is required and when it is required. Then, it is simply a matter of applying the contractually agreed process to set up the temporary transformer, feed it, and distribute from it. Or is it?

No, it is not. There is more to it. There is the diligence of managing this power supply equipment, and more than most any other common source of site service, this is the most critical. It requires maintenance. The leads in, the leads out, the contact points themselves, and so forth, require constant vigilance. Then, there is the need to be sure that no one is tapping into more than they are allocated. The best way to enforce this is to require those who need more than they asked for to bring generators for the additional power, at their cost.

Next comes the supply of the construction water. Although not usually as critical as the supply of the electric power, it also requires a plan and the monitoring thereof. The same goes for trash removal and dust control, two areas that, if not properly managed, will have effects similar to those from poor housekeeping.

Site facilities

Site facilities, such as offices, warehouses, prefabrication areas, changing rooms, and sanitary facilities, should be located to reduce the movement of personnel. The closer to the work that the office is located, the less time is spent by the supervision walking or driving from the office to the work area and back. If warehouses are remote, there is always the temptation for personnel to "need" an item that requires an unproductive half hour or so to get. Placing the changing rooms and sanitary facilities far from the work location is also counterproductive since the time required to go to and from the work site is not adding to productivity.

Prefabrication areas often cannot be placed near the work site, so when this is the case, thought should be given to fabricating smaller parts that are easier to transport over these larger distances. The cost may be less than the rental of the large equipment that often sits idle for days waiting for the larger pieces to be assembled.

Quality control and safety

Having said all of the above, it is of paramount importance that the quality of the work and the safety of the workers are never jeopardized. Although quality and safety will be addressed in separate chapters, they are briefly described here to emphasize the importance of making them a part of the site-management planning process, from the first day onward. When planning the flow of materials, equipment, and personnel, there are many opportunities to build in both quality and safety practices.

The flow and storage of materials can be planned so that no rework is required due to damage in transit or from poor storage, such as storing equipment in poorly drained locations, which could result in rusting or other deterioration. The movement of the equipment should be planned in accordance with personnel traffic patterns, to avoid loads being moved above the workers. A well-planned lift using a crane is shown in figure 7–1.

Fig. 7–1. A safely planned lift—there are no personnel below the load. (Courtesy of Construction Business Associates, LLC)

The personnel traffic patterns should also be addressed with safety as a foremost consideration. Routing of workers, especially at shift change, should be planned away from other ongoing activities such as heavy equipment movement, excavation work, and overhead construction. But an area frequently overlooked is the vehicular traffic in the parking lots. If the workers drive to the job or use a common parking area, it is important to design the flow of their vehicles with safety in mind. Possibly, a traffic signal may be required, especially where they leave the lot and enter the public thoroughfare. Sometimes, a flag person is required to control the traffic inside the parking lot, and often, physical barriers are needed to separate the workers walking to and from their vehicles from those driving into or out of the lot. Finally, when designing the traffic pattern for heavy equipment, barricading may be required to keep workers out of the path of the machine as well as the load it may be carrying.

When planning the site-housekeeping activities, thought must be given to waste material storage areas. If the material is flammable, it must be located so that no collateral damage could be incurred in the event of fire. If it is hazardous or dangerous to the touch, such as a sharp object, it may need to be barricaded.

While setting up and coordinating the site services, again safety must be a top priority. Providing and distributing temporary power is fraught with danger. Cables and wiring must be installed and terminated by licensed personnel. A third-party check should be made to ensure the safety of the installations, and periodic safety inspections as the job progresses will help to eliminate unsafe conditions that frequently occur due to weather, wear and tear, and even unauthorized personnel making modifications.

Another hazard that is often created comes from watering the roads on greenfield construction sites for dust control. If not properly scheduled, the roads can turn into slippery surfaces during the height of vehicular traffic. That can create a host of problems from vehicles sliding off the road to collisions between vehicles and construction equipment to personnel slipping and falling.

And finally, safety must be designed into the placement of the temporary site facilities, such as the offices, warehouses, prefab areas, changing rooms, and sanitary facilities. They should be located such that access to them is via safe, secured routes.

Site Supply-Chain Management

On any project other than a very small repair job, all manner of purchasing is done at the site level. This ranges from the owner, who may be purchasing consumables and even small tools for all site contractors to use, to the individual contractors, who will be purchasing whatever the owner is not supplying, to the service providers, such as the sanitary contractor who services the portable toilet facilities. The amount of money that is spent on a large, new construction project for these supplies and tools is tremendous. There are projects where these costs exceed 20% of the total field costs; they may reach millions of dollars. So properly setting up this area of the job is an important consideration for the site manager.

Unless there are existing purchase agreements already established for different jobs, often called "blanket agreements," it is usually the responsibility of the site management to make the arrangements for purchasing the supplies and services needed to support the site activities. Often, site managers are not well versed in the art of purchasing. Their background is managing labor; sequencing erection activities; and setting up rigging, machining, and welding activities. They operate under the assumption that whatever purchasing process worked at the last job should work here as well. Unfortunately, this is not true in the real world.

Site purchasing is a process that can have a significant impact on the financial results of a project. There are many different ways to approach this process, and each one has its pluses and minuses. For example, a large company can use its size to gain discounts due to corporate quantity purchasing. A local company can use its community presence to gain preferential treatment, and a company with an alliance partnership with the supplier can leverage future business opportunities to satisfy the needs of the current job. But whatever relationship is used, there is a need to define the roles and responsibilities required to make the relationship successful. Assuming that the supplier will provide certain goods and services based on a standard set of specifications (instead of tailor-made for the project in question) will lead to problems. It must be recognized that contracting to suppliers involves more than just price. It also involves the specifics of the service and delivery. But unless the site supervision has worked in the area before, or unless the site staff already has a relationship with the proposed supplier, there needs to be a formal process to ensure that the purchasing procedures are workable and satisfactory to all parties.

Especially during major outages with tight deadlines, failures in the purchasing process are seldom recognized until it is too late. The focus is usually on the end date, not the mechanics of the process or the efficiencies thereof. This can lead to an enormous amount of waste, on the order of 20% of the contracted goods and services. Then it translates directly into loss of profit, sometimes for both the purchaser and the supplier. Therefore, it is important to clearly define the performance expected, in addition to the price, when entering into agreements for the purchase of the goods and services to be used at the site.

Performance expectations

Ultimately, just as with the main construction contract, subsupplier contracts or purchase agreements come down to the same thing: Who is providing what for how much? When developing the service agreement, the overall job parameters must be placed as the paramount requirements. Many things can change as the project moves forward, from outright cancellation to changes in scope, and from changes in schedule—like acceleration or deceleration—to changes in site access and storage capacity. The more flexible the agreement is, the easier it will be to manage. Roles and responsibilities need to be addressed as explicitly as possible to avoid reopening negotiations due to these types of changes. During the heat of the project, it becomes very difficult to renegotiate about accountability that falls into unclear areas. If the service providers feel that their profitability is being impacted, they may use their position to leverage the deal, which usually results in an unhappy relationship for the remainder of the project.

Therefore, performance expectations need to be clearly defined and documented. Specifically, performance metrics and accountabilities should be established. As an example, if the order is for the supply of radiographic services, the agreement should spell out exactly how the service is to be performed, in accordance with specified codes or other criteria, and the skill level of the personnel performing the work. Additionally, response time needs to be established such as, "Personnel to be on-site within four hours of initial telephonic notification." Responsibility for barricades must be clear, and the time for delivering the film or digital data interpretation should be such that production work is not delayed. Also, final interpretation for weld acceptance should be clear, and the responsibility for retention of the films and interpretation records must be understood.

At a minimum, the typical site service order should include answers to the following questions:

- Who is accountable for enforcing the provider's commitments?
- What are those commitments?
- How are those commitments measured?
- What are the time frames for supply and removal?
- How often is the process reviewed for compliance with the commitments?
- What are the consequences for failure to meet the commitments?

The reality

Problems do occur, and in the heat of the battle, they are often blown out of proportion to the detriment of the job. Therefore, when structuring the service agreement, it behooves both parties to think through the process and plan at a detailed level. How each party will perform through each step of the process must be established and documented. Scenarios for both normal and abnormal situations should be reviewed. Steps should be developed to handle the abnormal situations, so as not to disrupt the ongoing flow of the portion of the job that is not impacted.

For example, suppose there are ongoing, six-hour nightly radiography sessions designed to keep up with production welding. Suddenly there is a need for radiography on a main steam line weld for a different contractor. The initial reaction from the radiographer may be to pull some of the resources from the regular work sessions to cover the steam line. But that may put a hold on some of the original production welding—an unacceptable situation. The service order must be flexible enough to adapt to these kinds of occurrences, avoiding the sacrifice of one part of the project for the other.

The resolution

Typically, the tendency of each party in a new construction or outage project is to look at its own bottom line, in isolation from the rest of the participants. But the project is not just about the individual players. It is about the investors—those persons or institutions that have invested their resources in the company that owns the project and who may pull out whenever the return on their investment is below expectations. Therefore, it behooves both the buyer and the service provider to look at synergy by becoming partners in the supply chain as opposed to unrelated auction participants. In other words, how can joining forces become more cost-effective than the usual purchaser–supplier relationship? By becoming partners, especially if there are financial incentives for joint efficiencies, the stage is set for ferreting out opportunities to create value during all phases of the work.

Although most company procedures require the typical vendor selection process to follow a "three quote and select the lowest bidder" scenario, that is exactly what often drives the relationship to be adversarial. An alternate approach is to use the bidding process only for identifying and prequalifying the suppliers. Then, the next step would be geared to maximizing value creation, as opposed to reducing costs through squeezing supplier margins and scope. However, to enter into a search for mutual value creation requires an understanding of each party's drivers and finding ways to achieve fair resolutions to common issues.

For the purchaser, these objectives usually are:

- On-time delivery
- Reliability
- Quality
- Responsiveness
- Technical capabilities
- Track record
- Financial strength
- Safety record

While for the supplier, the objectives are more like the following:

- Profit margin
- Order size
- Order repeatability
- Standardization
- Pricing arrangements (and cash flow predictability)
- Accuracy of specifications

- Delivery schedule
- Purchaser creditworthiness

However, to arrive at this mutually rewarding relationship requires open communications, both during negotiations and during the execution of the work.

> Investing the time and effort to understand each other's objectives and key decision-making criteria through open and timely communication provides greater transparency and increases the likelihood of developing a strategic supply-chain relationship, as opposed to a one-off purchase with limited long-term value.[1]

When purchasers identify suppliers with whom they have had successful relationships, they start looking at them as "key suppliers." Once perceived as key suppliers, these vendors have gained the loyalty of their purchasers and generally have the inside track for follow on business. If they have performed admirably during the construction or outage at site, whether for the owner or the general contractor, they then have the opportunity to be considered for a long-term relationship with the owner or plant operator, after the site activities are over, thereby generating even more business for themselves.[2]

Summary

Setting up the job site is a task that should be well planned. One way to do this is to prepare a project execution plan, or PEP, specific to the activities that will occur during the construction phase of the project. This plan should be developed with attention to the details that will make the management of the site much easier. It should address all of the steps in the process of managing the site activities before the actual labor efforts begin.

The PEP should be developed with the intention of making the administration of the site processes streamlined and effective. It must address the fact that the core of the site activities—the craftsmen—must have their needs serviced. They must be supervised efficiently; their paychecks must be timely and accurate; and they must have the tools, facilities, and equipment to perform their tasks. On the other hand, the management requires a streamlined and efficient flow of information. They must be able to respond to issues in a fashion that allows preemptive decision making, and this must also be addressed in the PEP.

The design of the site activities is critical. All participants in the project, from the owner to the individual contractors, should become involved in developing the flow of personnel, materials, and equipment, and they should do it with quality and safety as foremost considerations. It is these up-front efforts that will greatly reduce the potential for inefficient and costly movement of personnel and double handling of materials and equipment once the project gets under way.

Site services and site facilities can only be established for the good of all if they participate in the development of them. While the contracts between the parties will establish the basic responsibilities, no contract document will cover every detail. Working as a team at the outset of the project and establishing who will provide what service and facility for whom will go a long way to facilitating a smooth working relationship once the site activities are in full swing.

Finally, most participants on the site will also be purchasing a variety of goods and services. The outlay for this can be very significant, and it is often poorly managed. Working with vendors early in the project stage (before the pressures of the daily site activities reach high levels) can help to establish efficient flows of these goods and services. More importantly, plans can be made to handle abnormal situations that invariably will arise.

Careful planning, teamwork, and a cooperative attitude at the up-front stage of the site setup will go a long way toward ensuring that the project runs smoothly and effectively.

References

1 Budoff, S., and Krupinsky, V. (Saw Mill Capital). "Supply Chain Alliances Offer Added Value for the Power Industry." *Power Engineering*, November 2003.

2 Ibid.

Ensuring Quality 8

Often, one hears phrases such as "a job worth doing is a job worth doing right" or "it's funny how there's always money to do it right the second time." These phrases are expressions of attitudes, and quality is an attitude. Quality is a commitment that begins with the top corporate officials and flows throughout the entire organization. Adherence to quality and dedicated implementation of quality processes will be effective only when top management insists upon them; quality cannot be made when the commitment does not exist! Today's work environment demands not only a safe workplace but also quality products, services, and activities to satisfy customers' needs. Quality definitely impacts cost: Positive quality generates a positive impact on the cost of the work, whereas negative quality generates a negative impact. When there is a lack of quality, the symptoms are construction defects and the costs of correction. The disease is the lack of quality control.

Take the case of inadequate quality control for welding work. Suppose that the welders were allowed to draw rod from the welding rod room, go to the location of the work, and just start welding. Without proper welding rod distribution controls in place, the welders could accidentally use the wrong welding rod, which could result in the welded joint being weaker than the rest of the tube or pipe, a situation that might not manifest itself for some years. Or without proper weld inspections, such as radiographic checks of the welds, defects could be present in the welded joint that would not be discovered until the unit hydrostatic test, possibly requiring expensive installation of scaffold for access plus all of the costs of performing the reweld.

Controlling the quality of the site work requires that a clear channel of communication be established among the workers, their supervisors, the designers, and the quality enforcers. A policy must be established that sets the rules to be followed. A quality plan must be developed for each project, sometimes even subprojects, following the rules of the quality policy. Individual responsibilities must be assigned, and accepted, for the quality plan to have any chance of success. There must be a system of controls that can be used to stop the work in the event of poor workmanship or incorrect construction, and finally, there must be a feedback process that analyzes results and points to suggestions for improvement.

The Cost of Poor Quality

But first, let's explore the reasons behind the focus on quality. Let's look at what quality really entails. It is more than just monitoring the work at hand, such as the strength and slump testing of the concrete, the plumbing of the steel, and the monitoring of the welding. It includes a look at the total construction process, from the preplanning discussed in chapter 1 to the final reporting discussed in chapter 12.

During the discussions on preplanning and planning, one of the key activities was the lessons learned exercise, which suggested that various participants, including those who are stakeholders in the project and some who are not, brainstorm the project from top to bottom. The idea was to encourage out-of-the-box thinking, interspersed with hands-on experience. Upon completion, various ideas would have been generated that should help the project handlers and managers avoid the pitfalls that many projects stumble into. This is the first quality control step.

Next is to look at how the project gets put together. At some point, very early in the process, the thought of ROI (return on investment) pops up. In other words, does it make sense to pursue this effort? What will be the payback? Answering these questions requires an understanding of how much the project will cost, and quality is an important factor in this determination. Put another way, what is the cost adder of poor quality?

In the industrial contracting world, there are some who believe that many contractors' poor quality practices add up to 20% to the cost of the project. There are others who argue that it is barely 5%. The point here is not to argue who is right and who is wrong. The point is that no one disputes that there is some poor quality and that that poor quality has a cost. But what is that cost? Let's look at this from a contractor's view, knowing that whatever the cost is to the contractor, the plant owner will ultimately pay not only for this cost but also the markup on it.

Figure 8–1 shows the data collected by the author from a variety of sources over a three-year period. Obviously, the contractors do not want the sourcing revealed, and that has been respected. However, except for some rounding, the numbers are real, and the statistics are eye opening.

Project	Margine Slippage		Repair/Rework			Estimating Error
A	$ 1,500,000	=	$ 900,000	Engineering	+	$ 600,000
B	$ 1,900,000	=	$ 400,000	Materials	+	$ 1,500,000
C	$ 525,000	=	$ 250,000	Tube Leaks	+	$ 275,000
D	$ 725,000	=	$ 275,000	Weld Repair	+	$ 450,000
Total	$ 4,650,000		$ 1,825,000			$ 2,825,000

Fig. 8–1. The cost of poor quality

Four projects were analyzed, and the loss of profit for each is shown. What is not shown is the impact to the total bottom line, but one can see that it is definitely more than zero. More interesting are the causes attributable to these "overruns":

- Engineering and material fabrication errors cost $900,000 and $400,000.

- Pressure part weld leaks (during hydro) cost $250,000.

- 18% weld rejects (during radiography) cost $275,000.

- But even more importantly, poor estimating—in both the home office and the field—cost a whopping $2,825,000!

Refer to chapters 2 and 3, where the emphasis is not just on estimating and budgeting but also on experience and accuracy. Quality starts at the beginning of the project, and if it is poor, it is going to cost—sometimes a lot.

What are some of the other causes that belong in the definition of poor quality? How about lack of day-to-day planning like inadequate protection from bad weather, poor equipment delivery logistics, or inadequate availability of consumables? This continues with more items such as insufficient site traffic planning and, although often not lumped into a quality category, late contractor releases and notifications. All needlessly add to the cost of the project—the cost of poor quality.

Communication

Controlling the quality of the work on a power plant construction site is all about people communicating. No one wants to do a poor job. No one wants the work to be shoddy, but it happens all the time, and the challenge faced by most site managers is how to get it right the first time and avoid costly rework. The place to start is in establishing an environment of communication. This is not the same as a chain of authority; that will still exist. An environment in which the workers, the supervisors, and the engineers or designers are comfortable communicating with each other must be encouraged. This is not an easy task. It requires personnel management, ego stroking, and forcefulness, all at the same time.

The workers must want to do a good job; they need to be motivated to have pride in their work and feel a sense of ownership in the finished product, whether it is the construction of a new plant or a repair in an existing one. Part of the motivation process needs to include a clear path of information flow, or communication, of what is expected. The craftsmen must have quality-critical technical information at hand to remove the guesswork. In addition to a written program that usually stays in the superintendents' offices, workers need to be able to access this quality-critical information at the site where the work will take place. Going back to the welder, one way of aiding this process is for the quality control or welding supervisor to physically color-code the tube and pipe end joints with the same color that the weld rod will be marked with. In this manner, if anyone, worker or superintendent, notices weld rod ends in the vicinity of the work that have different colors than the tube or pipe ends, the work can be halted temporarily and an investigation can be conducted.

By providing the welders with work processes to promote quality, such as the color coding discussed above, they are drawn into the success of the project and encouraged to aid in the process of controlling quality. The welders are encouraged to communicate by receiving information that will allow them to do the job correctly the first time and to notify others in the event of incorrect work. The workers and the superintendent have the opportunity to prevent poor quality from happening in the first place by communicating with each other at the beginning of the work.

Similarly, communications between the quality control staff and the site supervisors must be open. The site supervisors are usually consumed by the day-to-day operations of the site, keeping the project on schedule, keeping the materials flowing to the workers, and providing them with the tools they need to do their jobs. They often neglect to put quality issues at the forefront of their decision-making processes. Therefore, it is important that the quality control staff interface with those responsible for the production of the work. It is important that the quality control staff communicate the quality requirements of the job. They must explain the importance of doing the work correctly the first time, and they must show what tools are available to facilitate this. The challenge for the quality control personnel is to convince the production staff that the effort to do the work right the first time is much less than the effort required to correct it later.

Then there are the communication channels that must be established between the site quality control staff and the off-site groups. There is usually a home office quality group that the site personnel must satisfy. The home office quality group usually sets the policies and procedures to be used at site. This group usually advises the site personnel on code and regulatory questions, maintains final interpretative responsibilities of quality tests like radiographs and calibration data, and also manages the audit processes. Without a clear communications channel, the site personnel can end up waiting for extended periods for responses from the home office group. Hold-points can turn into schedule delays, especially if third-party inspections that the home office is responsible to arrange are not timely. Imagine the need to get the insurance company's authorized welding inspector to witness the hydrotest, but no arrangements are in place. By the time this gets arranged, maybe three or four days later, the job could be subject to liquidated damages (LDs) for missing a critical milestone. For information on codes, standards, and regulations, please see chapter 4.

Planning

Setting up the job-site quality group requires some forethought. What will be the group's primary functions? Will the group be required to inspect off-site materials before they are shipped to the site? For example, if duct pieces are manufactured and assembled elsewhere, especially if done in a shop not accustomed to providing parts for power plants, will someone from the site be required to visit the shop and verify that the parts are being manufactured and assembled in accordance with specifications? Take the case of duct being assembled in a developing country where quality control, in the classic sense, is not the norm. Left unchecked, this material may well arrive at site with missing welds, out-of-tolerance squareness or roundness, and even dimensional defects. The cost to repair these types of errors on-site can be tremendous compared to fixing or eliminating them while the material is still at the shop. Remember the $400,000 materials rework in figure 8–1.

What about forgeries, especially for imported items such as high-strength bolts for the structural steel or even weld rod? How will the quality group members know if this is an issue on their job? The same goes for electrical parts, pipe pieces, and even rigging gear. Will the quality group be responsible to track and provide releases for work to continue, after hold-point inspections? Will it be held accountable if releases are not made in a timely manner due to an inadequate number of inspectors? How does it handle a sudden increase in defective work, which then requires a sudden increase in inspections? And what about all the record keeping, the paperwork, and the electronic data manipulations? The quality group usually holds the records that verify the acceptance of completed portions of the work, and it must be able to make these records available. When the site quality group is being established, questions like these must be asked and answered.

There will be someone appointed as the leader of the group. This person will have to work in concert with the site manager and possibly with other site organizations when determining how to structure the group. Some site managers view the quality group as an annoyance, both in terms of cost to the job and in terms of disruption to productivity, reinforcing the earlier attitude of "funny how there's always money to do it right the second time." Having adequate staff to do it right the first time is low-cost insurance. There is no need to let quality deteriorate to a point where rework affects the project program and where the cost of this poor quality affects the job-site budget, all due to not wanting to spend a bit more money up-front.

Also, when staffing the quality group, forward-thinking managers look to the turnover phase of the work, that stage where the construction group starts releasing the installed plant to the commissioning group. This is a period of pending chaos. Deadlines are usually just around the corner. Resources are stretched because the construction team is trying to demobilize. But someone must still prepare the mountains of paperwork required to ensure the smooth turnover of each defined piece of plant, equipment, or area. These turnover packages, often consisting of numerous reports and other documents, can be very time-consuming to prepare. The review of them by the recipient can also require extra time, especially if they are incomplete or incorrect. So staffing of the quality group may actually need to increase as the project nears the completion of the construction phase, just to avoid delays in turning over plant which could then lead directly to delayed start up and subsequent LDs.

Responsibilities

Who has the responsibility for the quality of the work delivered to the end user? Generally, the contractor, or general contractor, is contractually bound to a certain level of workmanship, and this workmanship is under-written by a warranty. Since sometimes defective work, usually referred to as latent defects, may not be readily apparent upon turnover. The cost for repairing this work, especially if it is not discovered until after the contractor leaves the site, can be very high. So it is in the interest of the contractor's site manager to ensure that quality is a priority. Usually, an analysis of the extra costs to make repairs and reinforce quality checks will show that this cost is preferable to performing warranty work later. Site managers must understand their responsibility when it comes to doing it right the first time.

Their challenge, then, is to set up their quality organization with specif-ically assigned responsibilities. A point person, or manager of quality, is normally assigned to accept overall responsibility to ensure that the quality process is right for the project and functions as designed. This person should be experienced in the type of construction being performed, for example, familiar with the art of welding and welding processes if there is much welding to be performed. The point person should be well versed in alignment and micro-measurement procedures if major turbine and turbine rotor work is done on-site. The point person should be familiar with electrical standards and structural steel and civil works procedures when the project consists of building a new power plant or adding new pollution control equipment.

Usually, quality managers know specific individuals whom they are comfortable working with. These individuals have probably been members of the quality manager's team on a previous project, and they should understand the peculiarities of a power plant project. They will know the importance of the paperwork required, and they generally understand the impact of paperwork delays, impacts such as delayed turnovers that can lead to delayed completion and ultimate LDs. Quality managers must be comfortable with delegating responsibilities, which will allow them some time to interface with parties exterior to their groups.

To be effective at managing the quality process on any construction site, quality managers must have the time and knowledge required to "play politics." Since quality control is often perceived as undue interference by the craftsmen and superintendents, it is important for quality managers to gain the trust and cooperation of these personnel. Quality managers have to show them the importance of working together to achieve a quality project. They also have to work with their home office staff, providing the staff with the assurance that what is being done on-site is in keeping with company standards and contractual obligations. Quality managers have to solicit their staff's support when contractual interpretations are required, and they have to rely on their staff to keep the group updated with revisions to standards and drawings.

The rest of the job-site staff also must be held accountable for the work performed under their supervision. Since the craftsmen actually work for these superintendents, not the site quality staff, it is the super-intendents who must manage these workers so that their workmanship meets the quality requirements of the job. The quality staff is responsible for performing the tests necessary to check on the quality of the work and advise the workers' supervisors when the work is substandard. It is these supervisors' responsibility to take corrective actions and implement measures to avoid further deterioration of the work.

Although the on-site quality control group, along with the workers' supervisors, is the frontline enforcer of quality, everyone else associated with the job also has a responsibility to ensure that the work meets all the quality standards prescribed. Those with frontline responsibilities must be held accountable to have their workers perform in accordance with established job-site standards. But those not on the frontlines also have a responsibility to support the system. They may be the material controls people, who must have a system in place that gets the right parts to the right place at the right time, in the proper condition. They may be the schedulers, who should be responsible to advise the site staff of potential issues or relief of constraints as soon as they appear, so that the frontline staff will have the opportunity to be flexible for work-arounds. And the top-level supervision, the site managers, must accept their responsibility to support their quality staff in the performance of the staff's duties.

Audits

Quality control is often perceived as an ugly word. It is not an issue of quality; it is an issue of control. But somehow, the quality of the work must be controlled. Quality requirements are usually spelled out by contract, often referring to various codes, laws, and standards. Some are also unwritten, accepted industry standards. But until an enforcement process is put in place, there is no guarantee that the prescribed standards will be met. The client, the end user, and the regulatory authorities want assurances that the work is performed in accordance with these requirements. So in addition to the quality control processes performed at site, an audit system must be invoked.

Generally, there are two types of audits: internal (self-audits) and external (third-party) audits. Both types have the same goal, assessing whether the quality controls are being followed. The internal audit is performed by personnel from within the organization being audited, but not by those who are performing the activities that are being audited. The internal audit plan should be developed taking into consideration the status and importance of the activities and areas to be audited as well as the results of previous audits. This audit plan should be reviewed after each audit and updated if necessary. The audit results should be documented and analyzed for quality trends. Senior management should be informed of the results.

External audits, generally subcontracted to a third party, should be used to remove any possibility of bias that may occur during an internal audit. Most quality programs require periodic third-party audits for this very reason. As with the internal audit, a plan should be developed taking into consideration the status and importance of the activities and areas to be audited as well as the results of previous audits. However, the auditing team must be given the flexibility to go beyond the plan and investigate any other areas or processes that they believe may impact the work they are auditing.

After the completion of either of these audits, the company management responsible for the areas audited should review the audit results and implement corrective actions on deficiencies found during the audits. Follow-up actions should then be implemented to verify that the corrective actions were taken, along with documenting those results.

At site, the quality group often perceives audits as intrusions into their space, just as the frontline workers perceive quality control as an intrusion into their world. But without these controls, there is no independent verification that the work is being performed according to requirements, and that the possibility of warranty work being required has been reduced and/or eliminated. To make audits go smoothly, the site quality group should envision what outside auditors would look for and what paper trails they would want to verify, and the site quality group should make sure that these are in place. They should plan for the interruption that an external audit will create in their operations and have a plan in place to provide coverage in the event they are required to be away from their normal duty stations.

For example, a third-party audit of a large power plant construction project might take several days. At a minimum, the lead quality person, the manager of the group, would have to dedicate a significant portion of time to this audit. If the paperwork were not readily available, other quality staff personnel would also be needed to identify, find (sometimes create), and explain the various forms and compliance to procedures. This would take away from the time for them to perform their normal duties, say, inspecting welds or verifying coupling alignments, which would then delay production work on the site.

Since this only adds to the perception that quality control is a hindrance to the work, it is in the best interest of the project for the quality group to be well prepared to maintain coverage of their duties while being audited. An effective way to accomplish this is to agree upon a schedule of audit activities, some possibly being performed after normal working hours, to minimize job-site disruption. With a timeline established, the auditors will know who will be available when, they will know what they can review when and for how long, and they will be able to streamline their work as well.

There is a cost for audits. The site disruptions and the time the site personnel spend on behalf of the audits is time not directly benefiting the job. That is not to say the audit is not a benefit, because it is; however, these are hours when the on-site staff is not deriving a direct benefit. Then there is also the cost for the auditors themselves. Usually, the job site either pays for these audit costs directly, or the job site is allocated some portion of the home office overhead costs for performing third-party audits. They can be significant, especially since they are often conducted on an extended schedule, and they involve the travel and living expenses of the auditors. But these costs should be viewed as the cost of ensuring that there will be a lower likelihood of rework required as the project proceeds and that there is less chance of warranty work after the job is over.

Analysis

No discussion of quality is complete without addressing ways to improve the process. Since quality assurance and quality control are people-oriented activities, the process cannot be reduced to the ones and zeros of the electronic medium. People work and people inspect—they are not robots, programmed to perform the same task in the identically same way every time. Therefore, there will always be variations. The intent of analyzing these variations is to reduce them.

An important function of the site quality group is to maintain statistical records of the results of their inspections. Similar to measuring productivity for the purpose of increasing output, measuring quality is for the purpose of reducing defects. As an example, measuring the weld defect rate of the welders will identify which welder is experiencing unacceptable rates. Analogous to productivity, when a welder is making welds that are not acceptable, the welder's productivity is subpar or, put another way, unacceptable. The quality control inspector, when charged with the responsibility of tracking the reject rate of welders, is actually contributing to the management of job-site productivity and the overall results of the project.

Let's look at the example in figure 8–2. This is a superheater and reheater replacement job with just over 1,000 2-inch welds to be made in eight days. It requires 40 welders working two 10-hour shifts, with a productivity of three welds per shift each, to complete the job per schedule. Also, let's assume a labor cost to the contractor of $65.00 per man-hour.

Cost Impact of Weld Rejects

Required to Meet Schedule	
1,065 welds	
8 days	133 weld per day
2 shifts	67 welds per shift
10 hours per shift	
3 manhours per weld	3 welds per welder per shift
	requires
	20 welders per shift at
	$65.00 labor cost per one manhour equals
	$207,675 total labor cost, or
	$12,980 labor cost per shift
Actual Results	
Weld Reject Rate	
0% for	17 welders = 3 welds per welder per shift
33% for	3 welders = 2 welds per welder per shift
	totaling
	63 average welds per shift with
	17 shifts required for job completion equals
	1 extra shift required or
note: numbers not exact due to rounding	$12,980 in labor cost overrun

Fig. 8–2. Cost impact of weld rejects

Now, let's say that through normal nondestructive examination (NDE), three of these welders each have one weld rejected every day. If this were to go on for the total eight days of the job, the result would be the need to work one extra shift, which in the case of this example would add almost $13,000 to the job. However, by diligent analysis of statistics, the quality control inspector can bring this trend to the attention of the welders' supervisors, who, in turn, can make a change to avoid this trend continuing. If the change is made early enough, say, after the second or third day, there is the chance that the extra shift will not be necessary and the job will have avoided this $13,000 adder. This is one way to avoid the cost of poor quality!

Summary

A shift is delayed here, a milestone is missed there. Was it because of a poor attitude that led to poor quality? Many times, it was. There is a quantifiable cost for poor quality. It goes far beyond just the cost of the repair. If it delays the job, there are concurrent overhead costs that cannot be recovered for the contractor, and there are revenues that cannot be generated by the plant. If it triggers LDs, there may be penalties that wipe out any hope of the contractor making a profit. If it leads to warranty problems after the job is complete, both the plant and the contractor may incur major costs and downtime. Poor quality is costly.

Poor quality must be avoided. This requires strong commitment from the top. It requires a system of communications among all parties, the field, the site office, the home office, and any third parties such as regulatory agencies. Since the management of quality is often perceived as an intrusion and a hindrance to productivity, a successful quality program must have people with good personnel management skills. The workers must be convinced to want to do a good job. Their supervisors must be made to understand that quality checks are good for the work. The right expectations must be communicated, and the tools required to do a quality job must be made available.

In the same manner, communications between the site and the home office support groups are important as well. For example, when the site depends upon the home office for arranging third-party inspections, this has to go off without a hitch because it is usually a stop-work hold-point that can affect the schedule and even invoke LDs.

It is incumbent upon the top site management to organize a quality management group that can vouch for the quality of the work performed. The group must have qualified individuals accustomed to working with power plant equipment. But if the job is broad, like a grassroots new plant project, then the group may also need skills in managing the quality of civil work, steel erection, and electrical work. Since a large part of the quality group's efforts can affect progress and the potential for warranty work, they can have a significant impact on the costs of the project, so understaffing in this area is not wise. The group acts as an insurance policy protecting against financial losses.

For the quality plan to be effective, to be in compliance with the quality policy that underlies the job, a verification system is generally required. This is the purpose of the audit process. It is there to provide independent verification that quality control plans are being implemented. Although sometimes viewed as an intrusion into the work at the site, this process actually lends credence to the work of the quality staff and, ultimately, to the quality of the work performed by the craftsmen themselves.

There are many studies on the cost of nonconformances. They point to millions of dollars needlessly spent and an equal amount needlessly lost due to the inability to operate the plant. Included are delays associated with rectifying poor quality as the job is progressing. There are the costs of the rectifications themselves, and there are also the costs of lost revenue generation on the one hand and penalties on the other. Then there are the costs of latent defects, those warranty issues that only come to light after everyone has left the site. Those repairs and their associated downtime can be crippling. A good quality control program can help alleviate this.

But a good program can also enhance production. By measuring quality often and analyzing the results daily, trends can be observed. When these trends show that certain work continues to be below par, there is often time to make a change and save the day. The quality staff should be in constant communication with the frontline supervisors, providing data and trend analysis to help them reduce nonconformances and increase productivity. Avoiding an extra shift or two by being proactive with the job-site super-intendents has a positive impact on productivity; it is the attitude of doing it right the first time.

Safety 9

Often underrated, safety can be one of the most costly components of a job. According to the U.S. Bureau of Labor Statistics, over 30% of the nontransportation- and nonhomicide-related fatalities in the U.S. workplace are in the construction industry.[1] That is a lot of unnecessary and costly carnage, and it calculates to an 8.9 fatal work injury rate, or almost nine out of every 100,000 full-time equivalent construction workers. And most of the time, these fatalities are preventable.

Everyone knows that construction is complicated, especially when building, rebuilding, or repowering a power plant. The above numbers show that it is dangerous. In fact, working construction in a power plant is deemed to be working in one of the most challenging safety environments in the construction industry. There are all kinds of hazards:

- Falls from heights
- Trench collapses
- Scaffold collapses
- Electric shock and arc blasts
- Repetitive motion injuries

In the early days of power plant construction, just as in the early days of all other construction, safety did not garner the focus it does today. Everyone can remember the photo of about a dozen workers eating lunch while sitting on a beam high up on a skyscraper in New York City, circa 1932. They were not tied off, and they did not have any kind of safety equipment, not even a single hard hat. A lot has transpired since then, but a lot still needs to transpire.

We know that today's emphasis on safety is not just a moral obligation, it is also a financial one. Job-site accidents, whether fatal or not, are very costly. They are costly due to a host of issues ranging from the medical costs to the resultant insurance premium increases to the inefficiencies caused by the disruption to the workplace immediately after an accident. And they can also be costly due to the American legal system.

People have changed their view of accepting responsibility. There has been a major shift from accepting responsibility for one's own actions to placing the blame on others for not preventing oneself from getting hurt. This shift has occurred throughout the social fabric of our current-day culture, encompassing everything from simple gadgets to public services to major projects. Today, the individual is asking the corporate world to protect him or her from himself or herself. Courts are passing judgments and governments are passing legislation that the corporations must protect the individual and that the corporation's management may be held accountable. This boils down to equating safety with financial risk to the corporation and personal risk to its management. If the corporation does not do whatever may be necessary for the individual to safely use their product or service, or to safely work in the corporation's plant and facilities, then the corporation will pay and its management may go to jail!

There is an emphasis on training and protective devices for workers on company property because companies do not want to have to spend money for losses resulting from failures to act or operate in a safe manner, and their management does not want to go to jail. They know that the cheapest losses are the ones that never happen. The responsible corporation of today will do what it can to make the individual aware of a potential danger and will insulate the individual from a potential danger. It will educate its workers about potential dangers and train them to avoid and/or protect themselves from the danger. And the corporation will do all this because it's all about the money. But before we look at the actual impact to the bottom line, let's look at some safety metrics.

Safety Management Metrics

OSHA recordables

The Occupational Safety and Health Administration (OSHA), is the main U.S. federal entity charged with enforcement of safety and health legislation in the workplace. Various legal acts have been promulgated to enable OSHA to do this. As a part of this process, OSHA has mandated that certain records be maintained related to workplace incidents. One of these is known as the OSHA recordables log. It is a recording of every injury or illness that requires medical treatment more than simple first aid.

To provide meaningful guidance and a method for comparison, these injuries or illnesses, commonly called incidences, are then compared to the total man-hours worked during the job, during the year, and within the geographic or corporate entity where work is being performed. To provide consistency in reporting, these incidences and related man-hours are then compared to 100 workers working 40 hours per week for 50 weeks in one year, or the equivalent of 200,000 man-hours.

OSHA recordables rate =
number of incidences × 200,000 man-hours
divided by
total man-hours worked at the job, during the year, or within the geographic
or corporate entity where work is being performed

An example would be a construction project where, at the end of the job, 1,200,000 man-hours were expended, and there were 25 OSHA recordable incidences. Using the above formula, this would be calculated as follows:

OSHA recordables rate =
25 × 200,000 = 5,000,000
divided by 1,200,000 = 4.17

This, then is a statistic that can be used by anyone wanting to assess the safety record of a project or the company.

When compared on an annual basis, one can readily see whether or not a company's safety record is improving. For example, the above 4.17 OSHA recordables rate is representative of the typical contractor working in the power plant construction industry around the years 2004–2006. Today, this rate is significantly lower, hovering somewhere below 2.0, or more than a 50% improvement over a period of almost 10 years. Put another way, for every 200,000 man-hours worked in this industry, there now are fewer than 12 OSHA recordable incidences instead of the earlier 25.

Other recordables

Depending on the company and its safety culture, there are additional recordables that also are tracked. One of these is the lost time incidence rate. This is very similar to the above OSHA recordable rate, but is defined as the number of times a contractor employee is off from work due to an incidence.

Still another rate that some contractors track is the DART rate (days away/restricted or job transfer rate). That is basically a way of looking at loss of worker productivity due to a safety incidence.

EMR (experience modification rate)

At the beginning of this chapter, insurance premiums were mentioned. This referred to the premiums paid, or in the case of self-insured companies, the monies accrued, for workers' compensation insurance. The experience modification rate (EMR) is a number used to calculate this. Basically, the higher the EMR, the higher the cost, or need for accruals.

Various states within the United States have policies and laws that regulate and/or oversee the workers' compensation industry. One such group charged to do this is a state Compensation Insurance Rating Board (CIRB). This board usually develops experience modification factors for employers who have workers' compensation annual premiums of $5,000 or more. An experience modification factor adjusts an employer's premium to reflect the difference between the employer's loss experience and the average experience that is expected for its classification(s) and size. The modification factor places an emphasis on the number (frequency) of claims and (to a lesser extent) the severity of workplace accidents. If an employer has better experience than is expected for an average employer in the same industry with a similar payroll, the employer receives a premium credit. On the other hand, if the employer's experience is worse than the comparable average, the employer receives a premium debit. The ability of the employer to directly affect its premium in this manner serves as an incentive to control or eliminate workplace injuries.

In general, each year insurance carriers report to the calculating agency the company's class codes, payrolls, and losses for the last five years. The computing agency then uses three complete years of data, ending one year prior to the effective date of the rating period, to make its calculation. For example, a rating in 2010 typically would not use 2009 but would include 2008, 2007, and 2006 in the formula. This means that it can take several years before a good safety record wipes out the effects of one bad year.

If the company is at the industry average, its EMR is a 1.0. If its experience is 20% better than average, its EMR would be a 0.80. If it is 20% worse, it would be 1.20.

The Financial Impact of Safety

Although ensuring that good safety is practiced is first and foremost a moral obligation for any corporation, the corporation is *encouraged* to ensure its employees practice good safety through measures that directly affect its profitability. These measures are varied. They go from the clearly visible medical costs of injuries to the insurance premiums just discussed that reflect injury frequencies. They go from the hidden costs of accidents to the loss of funds for corporate growth and improvement (fig. 9–1). And the corporation is not the only one that loses. The workers themselves often also lose dollars—in addition to their misery, which cannot even begin to be quantified.

The Real Cost of Job Site Accidents

Visible Costs per OSHA Recordable
Medical Costs $35,000
Insurance Premiums Increase of 6.0% $10,800

"Hidden" Costs per OSHA Recordable
Lost Time Wages for one crew $2,700
Damage to Tools & Equipment $1,500
Decreased Workers' Efficiency of 25% for Next Day $1,500

Total, Per OSHA Recordable $51,000

Total for 50 OSHA Recordables **$2,550,000**

Loss of Business . . . . *Priceless!*

Assumptions:
Contractor with $350,000,000 annual revenues whose:
  Average Labor Cost is $210,000,000 with 3,500 Workers
  Average Annual Payroll is $130,000,000
  Workers' Comp Premium is 10% of payroll

OSHA Recordable Rate of 3.0 which equals 50 OSHA Recordable Accidents

OSHA Recordable Accident Costs = 70% of Total Costs

Fig. 9–1. The real cost of job site accidents

Determining the money spent for medical claims due to on-the-job injuries is a fairly straightforward process. Although costs vary greatly, depending on the nature of the injury, overall power plant construction injuries range from eye injuries of several hundred dollars each to soft tissue injuries (injuries to the back, shoulder, or knees) of tens of thousands of dollars or more. An overall average insurance claim for all injuries incurred, taken from a variety of sources including contractors, utilities, and industry publications during the past five years, is in the neighborhood of $35,000 per injury. Obviously, there are contractors and owners who have a much lower experience, and there are others who experience double and triple these numbers. Also, these values represent medical costs only, not death-related payments, lost wages, or penalties imposed by regulatory agencies.

Insurance premiums are closely tied to these medical costs and accident rates. While referred to here as "insurance premiums," they also apply to those organizations that self-insure and are thereby required to accrue the money that will be needed for medical payments in the future. If a company's accident rate increases, so will its premiums (see previous section on EMR). And even though today's accident will not retroactively raise premiums, it will affect the rates in the future, which therefore increase the total cost impact of accidents.

The next cost category, the hidden cost of accidents, is not so easy to calculate, and therefore it is frequently overlooked. However, these costs are just as real, and they definitely affect the performance of the work and the bottom line of the job (see fig. 9–1). The following are some of these costs:

- Time lost by the injured worker, the supervisor now attending to the worker, and the rest of the crew, who get involved in helping the injured, or even if just standing there watching
- Damage to tools, equipment, and possibly installed materials
- Inefficiency due to a temporary decrease in the morale of the total job site

The cost of workplace accidents continues to spiral in ways almost impossible to calculate. For example, using the example in figure 9–1, the contractor generating $350,000,000 per year in revenues and experiencing 50 OSHA recordable accidents that year has suddenly incurred a cost of $2.5 million, or almost 1% of its annual intake. This money could have been put to better use in (1) safety training, (2) newer or better equipment for use by the workers, (3) hiring additional support staff, (4) providing additional payroll incentives or bonuses, or (5) returning more profits to the owners and investors. Or the contractor could have done some of each. Making these kinds of investments would probably return much more than a one-to-one ratio, something not possible if the money had to be used to pay out accident claims.

But the costs do not even stop with the lost opportunities for investment. If the accident results in an OSHA violation, penalties could range anywhere from a few hundred dollars to a million or more. If the accidents result in an increase in the EMR, then in addition to an increase in premiums, the contractor may also be subject to disqualification by utilities from bidding on future jobs. And if the ultimate disaster occurs, the loss of life, the contractor could easily be told to leave the job and not return for a long, long time.

In summary, one must recognize that profit lost through injury to workers is not recoverable from an insurance policy—it is unrealized money!

Communicating

Communication is the primary and most important tool available for managing safety. Management must talk with the supervision. Supervision must talk with the workers. The workers must talk with each other and their supervisors, and the supervisors, in turn, must provide feedback to management. There should be training, demonstrations, and regular support sessions that include the plant personnel and the contractor personnel, as well as union management when their members are on-site.

But for the communications to be meaningful, they must be based on a specific set of rules or standards. It is important that everyone communicates in like fashion, everyone talks the same language, and everyone follows the same plan. The best way for this to be accomplished is to develop a written safety program. Theodore Christensen of Liberty Mutual Loss Prevention put it this way:

> Everyone likes things in writing. It's a tangible guarantee that work will be done or rules will be followed. This is particularly true when it comes to developing and utilizing a consistent company safety

program. Putting together a written safety program is a critical first step towards documenting company policies and procedures for accident-free construction, yet some contractors still do not have one. It is much more than just a list of rigid work rules. Written safety programs are important because they delineate responsibilities and expectations for everyone. They also provide guidance for field supervisors so they can handle unexpected conditions. More importantly, a written safety program demonstrates a firm's commitment to ensuring employee welfare and building a better bottom line.[2]

But communication goes beyond just the written program. It also encompasses the supervisors and their responsibilities for implementing the safety policies. It includes the employees and their responsibilities. It includes management and its responsibility for providing the written safety program and the training to enable everyone to work within its confines. Then there are the external agencies, such as OSHA and its inspectors, who must be able to communicate with the parties being inspected. They must be able to discuss the latest legislated and administrated requirements that must be met on the job site and the governmental consequences of not being in compliance. And if there are any special incentive programs for the workers or for the supervisors, these must be clearly communicated so the programs create their intended results.

However, the most influential manner of communications is the direct one-on-one between the workers and between the workers and their supervisors. It is the responsibility of the supervisory staff to establish a teaming environment where workers will want to talk among themselves about safety issues and where they will also talk to their supervision about the same things. However, this requires that the supervisors have good interpersonal skills, something that is not always the case.

When supervisors are short on interpersonal skills, it is management's responsibility to provide them the training necessary to become good communicators. Ultimately, the supervisors must understand that the various parties involved in the project are individuals who have feelings that must be respected. It makes no difference if the individual is a first-time participant in the project or if he or she has been on-site many times before.

Managing the Process

Seldom are power plant construction projects performed in isolation. The owner/operator contracts a general contractor (GC). The GC contracts other major contractors, and they, in turn, subcontract specialty work. There are often multilevel tiers of contractors that will all be working on the same site, often on the same project and at the same time. It could be a formula for disaster, if not tempered with diligent planning and structured cooperation.

As already discussed, a lot is at stake when safety is not properly managed. The owner can lose, the contractors can lose, and employees can suffer serious injuries and die! When there are multiple levels of contractors, each with its own set of goals, rules, and responsibilities, the task to maintain a workable safety program can be daunting. How does the owner get its mandates imposed? How does the GC satisfy its client? How does the GC satisfy its own internal demands, and how does the GC ensure that the contractors below it stay in compliance with the owner's requirements and the GC's company's demands and still satisfy their own internal requirements? And what about the lower tier subs? They must work in compliance with the rules from all of the above. It is a delicate juggling act, but one that can be accomplished, successfully, if designed at the outset and enforced during execution.

From the owner or plant operator on down to each respective level, there are certain protocols that, if followed, will increase the likelihood of a cohesive site safety plan, whether the job is a three-year greenfield construction project or a two-week emergency plant turn-around. In general, five distinct, interrelated, but separate processes are involved: prequalification, contractor selection, prejob activities, work-in-progress activities, and evaluation (fig. 9–2). Since so much hinges on the successful integration of the safety operations of all contracting tiers, from saving lives to monetary savings to corporate survival, it is important to review the safety protocols.

Fig. 9–2. The contractor safety management process

Prequalification

First, stakeholder requirements must be understood. They must be spelled out and prioritized. They must include the requirements of all of the stakeholders "up the ladder," and often including the needs and requirements of others such as governmental agencies (e.g., OSHA) and special public demands. Once they are determined, these requirements should form the basis of a contractor prequalification questionnaire. This questionnaire should be the same for every tier of contractor to avoid misinterpretation of priorities once the job is under way. As a minimum, all tiers should be requesting the following information from their contractors and subs:

- Organizational hierarchy and authority
- Company work history
- Safety and health performance statistics including OSHA recordables and EMR for at least the past three years
- Regulatory citations for the past three years (OSHA, Environmental Protection Agency [EPA], etc.)
- Safety and health policies and programs
- Substance abuse program
- Insurance carriers and limits
- Safety and health training and evidence of use

Once the questionnaire is complete and satisfies the requirements of all of the stakeholders, it should be sent to all contractors that are interested in working on projects with the purchaser. They should be encouraged to complete the questionnaire in its entirety and also submit any additional information they feel will help the decision makers. Upon receipt of the completed questionnaires, they should be reviewed, and any questionable information should be clarified and/or verified. With this information now available, an evaluation can be made by comparing the information provided with the internal requirements of the company. This will then generate a list of prequalified contractors for consideration when planning future jobs.

Finally, it is important to ask all of the prequalified contractors to update their information annually. This is especially important for safety and health performance statistics and for regulatory citations. Also, the insurance information should be updated, and evidence of training should be provided every year.

Contractor selection

Once the prequalification process has been completed (i.e., the question-naires sent out, received back, and evaluated), a list will be available for use in selecting which contractors are to be invited to bid. Then, a selection process must be developed that will determine which contractors are suitable for what projects. Not all contractors will be suitable for every type of job; major differentiators in the power industry are fossil or hydro work versus nuclear, wind, or solar projects.

After determining which contractors are prequalified, the business needs of the purchaser enter into the decision of which contractors to short-list. The purchaser should prepare a bid specification that clearly spells out any specific safety requirements and any special conditions such as unusual hazards that may be encountered on the job. The special conditions might include existing or suspected asbestos or lead-based paint, cautions about underground utilities, warnings of expected vanadium on boiler tubes or arsenic in the fly ash, or information related to ambient conditions such as high noise areas. Contractors on the approved list of suppliers should then be interviewed to establish their ability to meet all of the unique require-ments of the job specifications. Those that meet the required criteria would then be selected to be asked to bid.

At this point, a prebid meeting will usually be held, with all of the potential bidders attending. At this meeting, all of the known and potential safety hazards should be discussed. The contractors should all be encouraged to ask questions and express any of their concerns. They should be told what the purchaser will be providing in the way of safety protections, and they should be told what they must price and provide. For example, if the owner or GC is providing first aid facilities and/or an on-site nurse, this should be stated. If the owner or GC is providing an ambulance, this should also be discussed. And if there is to be any cost sharing of these services, this information should be made clear during the prebid meeting so the contractors can prepare their price quotes accordingly.

After detailed bidding and negotiations, which at this point are more focused on technical and commercial issues, the safety element should be re-evaluated to ensure that compliance to the original requirements will still be met. If all is in order, a contractor can then be selected and an award made.

Prejob activities

With the contractor selection process completed, planning the job can begin. Issues such as site orientation, site-specific safety plans, site work plans, and training and compliance reporting need to be addressed. All parties should reconfirm who has the responsibility to provide which of the required safety services such as first aid, ambulance, and nurses. As with the contractor selection process, a lot of effort is required to properly plan for a safe and effective job, one that will allow the workers to go home the way they came, safe and sound. A cost-benefit process should be used, but the cost should not be that of the actions taken or the equipment purchased to implement a specific safety requirement, but rather the potential cost of not taking the action or not purchasing the equipment—in essence, the costs that could be encountered when luck is not on the side of the worker.

For example, some contractors will use distinct markings on hard hats or other clothing to identify persons new to the project. The intent here is encourage the current workers to recognize who is new and to look out for and help this new person work in a safely compliant manner. However, hard hats get scratched and scraped, and other clothing markers also become indistinguishable. Therefore, other contractors invest in distinctly different-colored vests for their new workers, deciding that the cost of these vests, even though higher than markings for hard hats or clothing, is a small price to pay in their march toward increasing safety awareness.

Next, the first step for all parties that will be involved in the project should be a thorough familiarization with the proposed site location, whether a grassroots location or an operating plant. The owner's rules and requirements should be clearly identified. This could mean that certain areas will be off-limits to workers during specific times of the day. It could mean that all personnel must be familiar with a particular emergency warning siren and the specific routes to be used in the event of an emergency. It may be that the owner is providing the workers' compensation insurance, and therefore any injured worker has to follow a unique procedure for treatment (more on this later).

Next would be a discussion of the actual planning of the work. Planning safety into a project is just as important as setting production schedules and planning for the delivery of equipment and materials. There is no substitute for thorough pretask safety planning. Each step of each owner's proposed work plan should have a safety element that addresses anticipated hazards and how to eliminate or guard against them.

For example, when preparing to lift heavy equipment like a waterwall panel or sections of a heat recovery steam generator (HRSG), the rigging and lifting work plan should include steps such as the following:

- Third-party review of the lifting calculations
- Accuracy of the weights involved in the lift
- Inspection of the lifting equipment (crane, tuggers, or jacks)
- Inspection of the wire rope for size and wear
- Clearance of the lifting path (e.g., no structural steel in the path of the lift)
- Clearance of the area of any nonauthorized personnel

Similar to an airplane pilot always going through his or her checklist, not following these procedures could have catastrophic consequences. Not having a third-party review of the lifting or rigging calculations can be disastrous. Notice the severely bent spreader beam at the top of the photo in figure 9–3. Fortunately, the only result is the bent panel. There were people standing nearby, and if the rigging had snapped, someone could have been injured.

Note the bent
spreader beam

Fig. 9–3. A third-party expert should review lifting or rigging calculations to avoid life-safety issues, not to mention damaged equipment. (Courtesy of Construction Business Associates LLC)

In the photo in figure 9–4, where a boiler steam drum is being raised, the area below the lifting operations is designated as off-limits with barricade tape to keep people out. This is a double protection so that if the lifting cable snapped during the lift, there would be less chance of injury to personnel. However, if the cable snapped and the drum fell back down, there would still be major problems for the whole site. There would be the cost of repairing any damage to the drum, lifting equipment, and any areas involved. Rerigging and raising the drum again would double the cost of that activity. Then there would be the lost time. There would also be the lost productivity, and finally, there would be a mood change with the workers wondering why basic safety issues were not being addressed.

Fig. 9-4. Barricade tape is an important safety requirement for any heavy rigging and lifting operations at a job site. (Courtesy of Foster Wheeler Corporation)

And finally, as part of the prejob activities, safety training should be established. Most projects will require an initial site-specific orientation for each and every worker who comes on-site. This orientation may take four hours or more. Some sites also require that before even being allowed on-site, each person, whether a worker, supervisor, or outside manager, must have successfully completed a 10-hour OSHA safety course. There also may be specific situations that require additional training, such as confined space access, scaffold building, forklift truck operations, fire watch, hazardous material handling, and so forth. All of these requirements should be reviewed prior to the actual start of the work.

Work-in-progress activities

Once the site work starts, continual monitoring by management of the safety process is very important. Contractors must monitor themselves, and owners and their GCs must monitor the contractors below them. There should be specific reporting requirements, stipulated in the contract, that every level of contractor on-site should be following, in addition to the regulatory OSHA requirements. Regular inspections should be made and documented by the contractor performing the work as well as by higher tier contractors and the owner. In addition to individual contractor daily inspections, a weekly composite inspection conducted by the owner or GC is useful, so that the different parties can see, firsthand, their impact upon one another.

As part of the inspection process, conformance to policies and procedures should be verified. If a contractor's procedure requires a medical profile of every employee on the job, it should be verified that this is being properly reviewed, noted and filed as required. If every employee is required to obtain and read an employee safety handbook, it should be verified that each employee has received one. Verification of foreperson and supervisor training should be available along with all training records and certificates. A spot-check of the correct use of the substance abuse procedures should be done, and a regular review of the accident-reporting process should also be performed.

In the unfortunate event of an accident, it is extremely important to investigate, immediately, the issues surrounding the accident. There should be a standard format for this investigation, and it must have enough information to satisfy the contractual requirements of the job, the internal requirements of the contractor, and the regulatory requirements of OSHA or the EPA. But in addition to these and other standard requirements, a section should be included to (a) describe what actions could have been taken to prevent the accident and (b) what action is recommended to prevent a reoccurrence of the incident. The same holds for near misses. They also need to be reported, investigated, and communicated just like real accidents.

As part of the work-in-progress safety management and resultant cost control, regular reporting must be done of safety walk-downs, specific inspections, and accident investigations. Owners and GCs should always keep detailed records of what their contractors are and are not doing. This will help in identifying issues that are unique to the site, in time to correct them before an accident occurs, and it will be helpful for updating their contractor prequalification records, in support of the next project.

For the contractors actually employing the workers, regular reports beyond just the regulatory OSHA logs should also be maintained. These reports should be designed in a way that they can be used to record safety results and manage safety trends. In other words, since often only what gets measured gets done, if safety trends are correctly measured, they can be managed.

Additionally, every job site should be prepared for emergency evacuations, especially in an already operating plant. When an emergency occurs, whether it is a weather-related event such as a tornado or an earthquake, or a riot, a coal bunker explosion, or a steam or gas line rupture, the people working in the area must know what to do and where to go. Predetermined exit routes should be established by the owner and explained to all employees on-site. This should be done during the site orientation and repeated at least monthly during safety meetings.

Finally, if safety trends are moving in an undesired direction, or if there is a serious incident or serious near miss, management should consider a job-site stand-down. The employees should be expected to follow predetermined exit routes to their meeting point. There, management could then address the issue precipitating the stand-down and reinforce the importance of working together for a safe workplace environment.

Evaluation

The final step in managing the safety process is closing the safety management loop (see fig. 9–2). The loop started with the prequalification of contractors, went on to their selection, then followed with a prejob activity phase, followed by a work-in-progress step, and now must be closed with the evaluation of the previously selected contractors.

If the proper records have been kept, updated, and evaluated for performance management, then the information exists to evaluate the contractors on their safety performance during the execution of the work. Depending on the criteria selected by the principals, each contractor could be compared to its required standards and also evaluated against the commitments made in its safety policies and programs. Records of site visits by the contractor's managers could be used to determine their commitment to safety, and if problems were encountered, their responses could be used to gauge the depth of their seriousness. These, then, would be the criteria used for the next round of prequalification evaluations.

Drug Testing and Some Other Peculiarities

Most construction projects, whether they are for work on a power plant or any other industrial facility will almost always require some sort of drug testing. Many require preaccess testing for first-time workers. Some also require random testing throughout the duration of the work. Almost all require it in the event of an accident or even a near miss.

According to the U.S. Department of Labor (among many other studies), illicit drug use and alcohol abuse are more prevalent among construction workers than workers in any other commercial industry. Over the years, investigations have shown that these abuses have been a significant contributor to construction accidents and near misses. Fortunately, the industry has risen to the occasion and taken steps, such as drug and alcohol testing, to combat this safety (and productivity) issue.

However, there is still a long way to go to eliminate this hazard from the workplace. The recent rebirth, albeit small at the moment, of new-build nuclear power plants has provided some insights into how serious this problem remains. According to a March 14, 2013, article in the *Augusta Chronicle*, workers at the Georgia Power Vogtle nuclear plant were given 8,744 tests in 2012. Of these, 5,440 were given to construction-related workers involved in the building of Units 3 and 4, and 101 positive results were obtained.[3] That means potentially there could have been 101 accidents (that were avoided only because of this testing).

In addition to more emphasis on drug testing, there is also an increased focus on proper reporting of worker injuries. According to various news reports, in April 2013, a contractor's safety manager was convicted of lying about worker injuries at some Tennessee Valley Authority (TVA) nuclear facilities. This allowed his company to (illegally) collect $2.5 million in safety bonuses. His conviction resulted in a 6.5-year prison sentence, and the company had to repay TVA twice the amount of the bonuses.

Finally, there are two more reasons why renewed vigilance is required. The first is the influx of foreign-born construction workers. According to the same U.S. Bureau of Labor Statistics report referenced at the beginning of this chapter, one-third of all construction fatalities befell foreign-born workers. This is due to a combination of language barriers and unfamiliarity with safe work practices. Unfortunately, many of these workers came from countries where worker safety is not a paramount corporate principle.

There are still countries where the mantra is that there is always someone waiting at the gate for a job vacated by an injured or killed worker. These people need extra support during their safety training and throughout their employment.

The second reason for renewed vigilance is that construction workers are working longer. Instead of retiring in their mid-40s to 50s, many are now staying in the trades well into their 60s, and some even beyond. Unfortunately, the aforementioned U.S. Bureau of Labor Statistics report shows that construction workers over 65 years old have three times the fatality rate than younger ones. So this poses new challenges for those charged with managing job-site safety.

Some Options

So far, most of this chapter has been devoted to traditional ways of working with safety. But these are not the only ways to manage the process. Today, the construction industry in general, and the power plant construction industry in particular, have reached a plateau in the quest to reduce safety incidences. Great strides were made during the past two decades, with some companies reducing their OSHA recordable incidences by factors of 25% or more *annually*. However, there comes that point when reducing by percentages is no longer realistic. For example, if a contractor has two recordable injuries in a calendar year, reducing this by 25% is not possible; it must be either 50% or 100%. Using the old method of the "carrot and stick" will not work when striving for reductions of 50% or 100% at a time. Different methods are needed. The ones that follow are not necessarily new in concept, but they have yet to be embraced by all of the industry. Maybe that will change, as more and more emphasis continues to be placed on managing the costs of the power plant construction process. As has been noted by various members of the industry's management, enhancing a company's safety performance will reduce costs and improve profitability.

Owner-controlled insurance program (OCIP)

When a power plant owner decides that the costs of managing the construction work at a plant or job site need a dramatic reduction, one of the most effective ways of doing so is to remove risks from the contractors and take them on directly with an owner-controlled insurance program (OCIP). This is what happens when the owner takes over the insurance responsibilities for a construction project. By doing so, the costs of the insurance premiums or accruals no longer reside with the contractors; the owner assumes them directly, often at a reduced rate, since they may be spread across a larger base. But what insurances would the owner control? Depending on the owner's propensity for assuming (and managing) risk, they could be any of the following:

- General liability insurance, which includes coverage for equipment and property damage, personal injury, third-party involvement, products and completed operations, and employer's liability.

- Excess general liability insurance, which provides coverage in excess of the limits of the typical general liability policy.

- Professional errors and omissions liability insurance, which could provide coverage for negligent acts and errors or omissions by those contractors whose scope of work includes providing design and other professional services.

- Pollution liability insurance, which provides coverage for liability insurance arising from pollution releases during construction work.

- Builder's risk insurance, which usually provides coverage under an "all risks" format for physical loss or damage to the work or any part thereof, generally also including floods and earthquakes. However, it frequently excludes coverage for loss of or damage to materials not being incorporated into the project and for the tools and equipment being used on the project.

- Workers' compensation insurance, which is the coverage required in the event a worker sustains injuries related to his or her performance of work on the project. Not all owners who provide projectwide insurances elect to provide workers' compensation coverage for the project, but when they do, the savings impact to the project is generally the largest.

The cost savings to the owner when implementing an OCIP varies depending on the coverages elected, job size and duration, job scopes involved, the track record of the owner in managing risks, and the contractors' history of risk control. However, a study performed by the author in the late 1990s while working at a Midwest utility showed that providing coverage for just the workers' compensation component of a series of plant renovation projects removed a whopping $5 per man-hour for each and every man-hour worked by the contractors that were selected to participate in the program. Today, that savings would be even greater.

Not all contractors were selected to participate because some contractors' work scopes involved work that had very high risk associated with it, like asbestos abatement and scaffold erection and dismantling. But for the majority of the contractors, removing $5 of cost for every man-hour they worked added up to significant savings to the utility, which then used some of these savings to actively manage the safety aspects of all of the construction activities across a multitude of plant sites. The accidents that did occur cost on average $3,000 each, but were very rare due to the active participation of the utility and all of its contractors in managing the safety process. As insurance rates continue to increase, this may be one of the most cost-effective ways to curtail construction costs, provided that the owner can manage the process.

For the work performed at these sites, several specific actions were taken by the owner. First, a rigid contractor prequalification and selection process was used. Then, safety teams were formed with the power station operating personnel integrated into these safety teams. The local union leadership was also included in all safety meetings and inspections. This enabled all parties to see the impact they were having on each other and fostered a team spirit that essentially eliminated finger pointing.

In addition to these moves, the owner also provided full-time, on-site, construction-trained nurses. Their responsibilities included collecting and processing the drug-testing samples of all of the workers when they first came on-site, during random testing, and in the event of an accident. They tended to any and all incidences of injury and, if necessary, arranged for off-site medical attention. They were also responsible for site-specific safety campaigns, from publishing weekly safety letters to providing attention-getting safety posters.

Members of owner management also took a heightened interest in how the safety process was being managed since the risk of the program was now squarely on their shoulders. They visited the work site frequently; they interacted with the workers, the supervisors, and the safety teams;

and they arranged special forums to demonstrate to all of the participants that the owner was *very interested* in having a safe and healthful project for all. The bottom line result was a positive monetary savings for the owner, when measured against previous projects, and a significant reduction in the OSHA recordable rates for the project overall and for the participating contractors as well. So not only did the owner win, but so did the contractors.

In addition, all of the foregoing can also be accomplished by GCs on large, long-term construction projects, by using a contractor-controlled insurance program (CCIP). Moreover, OCIPs and CCIPs are not restricted to single projects. They can be used to cover a portfolio of projects over long periods of time. OCIPs and CCIPs are commonly referred to as wrap-ups because they wrap up four critical project components: insurance, claims management, safety, and risk control.

Behavioral-based safety

No discussion of safety would be complete without discussing behavioral-based safety. In a nutshell, the principle of behavioral-based safety is to get the workers to take responsibility for themselves by fostering an attitude of safe practices. The emphasis is shifted from focusing on incidences that have happened to incidences that were avoided.

Behavioral-based safety is a process that is designed to make the workers aware that *they* are the first line of defense for their own safety. Its underlying tenet is that the root cause of unsafe behaviors is that individuals' attitudes, beliefs, and values place time, comfort, and convenience ahead of safety. The belief is that there are two basic types of behavior that cause accidents: (1) unconscious behaviors, characterized by daydreaming, inattention, and repetitive tasks, and (2) conscious or deliberate behaviors, characterized by taking shortcuts or exhibiting other risky behavior.

Behavioral-based safety programs are not easy to implement. Their underlying premise is that attitudes must be changed, and then the new safety-conscious attitudes must be constantly reinforced. This takes time, and many power plant outage jobs are not long in duration. Usually, when the craftsmen arrive at the job site, they are immediately assigned to specific tasks with the weight of the schedule bearing down on them. Time is usually of the essence, and unless the job is a long-term, new construction project, no one takes the time to assess workers' attitudes, indoctrinate them into a new mind-set, and then constantly monitor their behaviors. However, to escape from the plateau where many contractors (and often owners as well) have been working, this attitude has to be revised.

Behavioral-based safety, as the name suggests, is a system of reinforcing good safety habits, as opposed to the usual system of measuring bad safety results and implementing procedures to reduce the statistics. A typical method of reinforcing good safety habits is to develop a list of observable behaviors that can be tracked and trended. For example, a typical list might include the following:

- Wearing hard hats
- Wearing safety glasses
- Using face shields
- Using burning goggles
- Wearing hearing protection
- Wearing safety harnesses
- Identifying new personnel

The employees are then observed, maybe twice per shift, and their behavior is recorded—how many are wearing or using the above protection devices and how many are not. These data are then presented as a percentage of safe acts with the target being to reach 100%. The goal is to have everyone focus on safe acts by providing feedback to them of their safe behaviors.

The old method of continuing to focus on unsafe acts, once the safety goals have been reached, loses its effectiveness because there are fewer unsafe acts to record, and with fewer acts to record, there is less data for feedback. Less feedback, then, means less focus, and less focus evolves to less effort to go beyond the plateau.

The ideal approach to implementing behavioral-based safety is to start long before the site work actually begins. Workers' attitude assessment and behavior modification training should start some weeks or months before. With a unionized workforce, or with a permanent plant team, training could be planned and implemented in a manner to coincide with the start of the project. Then the project would serve as a continuing training ground. Although this approach appears to ask the parties to spend money for safety training before the site work even begins, this should be balanced against the reduction of the cost of safety incidences that will occur once the work begins; it should be viewed as the cost of avoiding costs that could be incurred by continuing to use the old methods.

When it is not feasible to start the attitude assessment and behavior modification before a job begins, there are still ways to implement this process. There are various studies that show that 80% of workplace injuries result from the actions or inactions of 20% of the workers. This suggests that by focusing on those 20%, first and foremost, the majority of the unsafe behaviors usually found on the job site can be arrested and revised before they turn into accidents.

For example, the previously discussed safe habit monitoring could be increased to every two hours for the first day or two of the job. Very likely, there will be a small percentage of workers who will be found not using their personal protection equipment (PPE). These individuals could then be targeted for closer observance than the remaining workforce, and they could also be singled out for additional safety training and behavior modification.

Let's take an example. Figure 9–5 is a representation of potential safety behaviors and their related costs superimposed on an behavior scale. The bottom half of the chart depicts typical unsafe behaviors and the potential costs to the organization. For example, not wearing safety glasses could cost anywhere from nothing to $10,000, depending on the severity of the accident that would have been avoided had the proper glasses been worn. The middle of the negative behavior scale suggests that costs of anywhere from $100,000 to $500,000 could be incurred if a worker neglects to fix an unsafe area, such as replacing a piece of grating that he or she sees has been removed by someone else. Then there is the drastic bottom end of the scale, showing the potential disaster that can result if a worker neglects to lock out a piece of equipment before working on it—*death*.

But what about changing attitudes? With appropriate attitude adjustments, behaviors change. The upper half of the scale in figure 9–5 demonstrates the cost avoidance, or savings, that could be expected when positive behaviors are demonstrated. In the case of this example, a worker speaking to co-workers and suggesting that they put on their safety glasses, could result in avoiding safety-related costs of up to $5,000. By speaking to a supervisor about unsafe situations, such as noticing that a handrail is missing, or that a rigging cable is frayed, could avoid major safety-related costs, possible up to a $500,000. And finally, the sky is the limit when it comes to cost avoidance resulting from group discussions—the old adage of two heads are better than one. Group discussions can take place during safety meetings, during lunch breaks, during training sessions, and even when off the job. The key here: *communicate*.

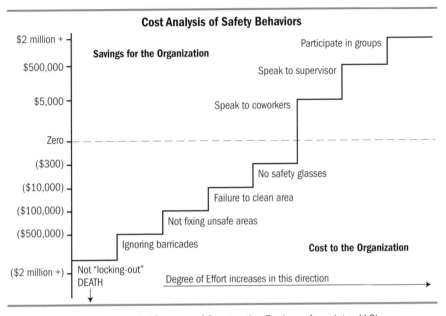

Fig. 9–5. Safety behavior scale (Courtesy of Construction Business Associates LLC)

However, to reinforce this positive safety behavior, some type of measuring system must be developed. Contrary to the typical system used for tracking bad safety habits, this system must focus on good safety habits. The idea is that feedback and recognition of positive safety behavior will feed upon itself and create a snowball effect of working in a safe manner.

Zero injuries

One of the elusive goals of all construction safety managers has been: How do we achieve zero injuries? The implication is that by eliminating all at-risk behavior, zero injuries will result. There certainly are organizations that go for an entire year with not one single injury; however, most companies still follow the mantra of comparing themselves with the industry averages, whatever they are. The industry averages, of course, are derived from a pool of the results of many companies and can range anywhere from very high numbers to the elusive zero. So when comparing a company average to the industry average, it is not possible to determine which is a consistent zero injury company and which is not—it is not even possible to determine which is a good safety risk. It is only possible to say that the numerical average of a company's safety results is better (or worse) than that of its competitors—whatever that means.

So the concept of zero injuries is often talked about but seldom pursued, even though there are some organizations that regularly achieve it. And why is that? Why settle for the average? Why not emulate the best? There is research that suggests that the return on investment for implementing a zero injury process can be as high as 300%!

Similar to the behavioral-based safety philosophy described earlier in this chapter, zero injury management requires a change in culture. It requires a change in the thinking of management and employees. It requires a change in the belief that injuries are inevitable; it requires a belief that although injuries do occur, this does not mean that they must occur! It requires a major training effort to reprogram the thought process of most of the employees as well as their managers.

One of the obstacles in attempting to instill a zero injury concept into the thought process of management is the old numbers game. Numerical goals are set that challenge the previous reporting period, but these goals are still numbers higher than zero. There is also the belief that zero injuries are impossible; however, there are organizations that achieve this, although not many. And then there are those who say that the ones with the low numbers or the zero rates are just plain manipulating the statistics—that figures can lie and liars can figure. The zero injury concept says that setting goals of anything more than zero sends the message that *some* injuries are acceptable!

Research has shown that there is always something else, or something more, that can be done to be safer. Sometimes it is mechanical, such as additional guards on a piece of machinery or barricading certain areas, but other times it is a process of training and educating workers to think of safety *all the time*. Maybe the time to start is today. One often hears that zero injuries are impossible, but one only needs to stop and think— zero injuries are impossible over what period of time? Did an injury occur today? If not, then maybe the approach should be to simply repeat today's actions tomorrow, and the day after tomorrow, and so on into the future.

Another way to approach the skeptics of zero injury goals: Start a project by saying that there is a new program to significantly reduce injuries on the job, say, only two OSHA recordables. Then look at the skeptics and ask them which two workers should be assigned to be the injured parties, and have them go and tell those two workers that they are it—they have been targeted as the sacrificial goats.

The zero injury concept requires a commitment to sacrifice other project goals in favor of safety. Schedules may have to be compromised, and productivity may take a hit but when doing the math, running the numbers to see what a serious injury or death would cost versus some liquidated damages for delays or some additional staffing to countermand productivity issues, working safely is usually the more cost-effective approach. In other words, safety must become a core value within the organization for zero injuries to be a reality. And once it becomes a core value, it *will* contribute to the bottom line of the organization.

But what needs to be done at a site level to support a zero injury process? Often, it not much different from what has always been done:

- Prejob activities, as described before
- Safety orientation and training
- Written safety policies, as discussed earlier
- Written safety programs, so everyone is following the same process
- Worker involvement, including the family
- A constant reminder, such as awards, free lunches, or incentive programs
- Substance abuse programs
- Accident investigations, including near misses
- And most of all, demonstrated commitment to safety by top management

The site managers must mirror their top management's position that the company has a commitment to zero injuries. There will be no safety goal other than the number zero. There will not be any comparisons to anyone else. There will be just one simple measurement—there were injuries or there were not—no more, no less. There will be a complete devotion to the elimination of unsafe behaviors by all employees, workers, and management alike.

However, care must be taken that when zero injuries are reported, zero injuries are actually sustained. It is common for employees to rally behind a corporate cause solely because it is expected, even though not realized. Employees may hide some injuries, and this is not the intent of the program. Hiding injuries does not translate to better productivity; it does the exact opposite, the worker is now less productive and the intended cost savings are not realized. When working toward a zero injury result, any injuries that do occur must be acknowledged so the cause of the injury can be investigated and rectified, or else it will happen again and again, negating the whole process. The zero injury safety process is powerful when properly implemented.

A final word about hiding injuries: If an injury, no matter how minor, is not reported immediately and later develops into something more serious, the insurance carrier may refuse payment for medical and/or compensation costs. For example, a minor contusion injury to the neck or shoulder is not reported. Weeks later, it develops into a blood clot that requires hospitalization and surgery. These costs would be denied by the insurance carrier because there was no record of the injury that led to the blood clot. And even if fellow employees insisted that they witnessed the incident, since there was no official record of the injury occurring, no payment would be made—how could the insurance carrier be sure that the incident did not happen while the employee was away from the job, unless there was a record on the job?

Workers' compensation fraud

All site management—the supervisors and the foremen—should understand the workers' compensation law and how much a continuing claim can cost the company. Since the supervisors are usually the managers who have the closest working relationship with the workers, their support is key to making the process work. In the event of an injury, the supervisor should get in touch with the employee and his or her attending physician. They should let the employee know how much everyone looks forward to his or her return and how valued the employee's contribution is.

The supervisor should talk with the physician and ask for the employee to be released for alternate duties as soon as possible, and to advise which tasks are not suitable for the employee to perform. Preferably, the employee should be returned to his or her original job, perhaps with restrictions on activities. Otherwise, an alternative job should be found. Either way, the worker should be returned to duty as soon as possible and the job assignment should be upgraded as the worker recovers.

Managing the return-to-work process should not be the duty of one individual supervisor, but should be shared among the supervisory team. Often, it should be done in concert with the insurance company, which usually has much experience in helping a worker to full recovery. The process should be regarded as a financial management challenge to obtain the maximum productivity from the employee while keeping the medical expenses to a minimum. This requires close communication among the employee, the supervisors, the physicians, and the insurer.

However, an unpleasant but realistic possibility that must always be considered is fraud. Unfortunately, fraudulent workers' compensation claims occur, such as claims where the worker was not injured on the job, claims where the injury is an aggravation of a previous ailment, and claims where the severity of the injury is exaggerated. These kinds of claims are very costly to the insurance and construction industries. Fortunately, there are some indicators that the site supervision can use to monitor for the possibility of fraud:

- Disgruntled employee was facing firing or lay-off.
- Employee has short-term employment history.
- Employee is experiencing financial difficulties.
- Injury occurred early Monday or late Friday.
- The injury was not witnessed.
- Details of accident are vague.
- After the injury, the employee is seldom home.

Good communication between the supervisors and the worker and between the worker and the worker's peers will usually ferret out the above issues, which can then be addressed with the worker and the insurance carrier. It is important that all alleged injuries be challenged; insurance carriers employ investigators for just this reason. Since fraudulent claims are usually high dollar claims, if they are not dealt with promptly and properly, they could go on for years.

Summary

Accidents result in unrealized profits. There are the direct costs, there are the hidden costs, and there are the costs of lost business. There is also the possibility of damaged worker morale, regulatory penalties, and jail time for the executives. And none of these costs are recoverable through an insurance policy.

But what are accidents, and how are they measured? In the United States, there are federal laws mandating certain actions to be taken by any company involved in the workplace, including the construction industry, to protect its workers. Not following these laws can become very expensive, very quickly. In addition, these laws, enforced under the auspices of OSHA, require certain reporting. The most common of these reports is the OSHA recordables log. This is a log of safety incidences compared to a common denominator of the equivalent of 100 workers working 40 hours per week for 50 weeks in one year, or the equivalent of 200,000 man-hours. An incidence is any injury or illness requiring medical treatment more than simple first aid.

How important is good safety management to the bottom line of a project or a corporation? While the actual medical costs of an injury can be very high, there are a host of ancillary costs that can be even higher. For example, there are the costs associated with lost productivity of those involved in and around the accident. There are the costs of repair in the event of damage. There are the costs of higher future insurance premiums. And depending on the severity and frequency of accidents, there can be the costs of being terminated on the project and/or not being allowed to participate on future projects. So how is all of this prevented?

The first line of defense against job-site accidents is proper communication. The needs of the stakeholders must be communicated to the organization, which in turn then puts these needs into written policies. From the written policies, written safety programs can be developed, and these programs can then be used as tools to manage the safety process. For details on how to develop and prepare these policies and programs, please see the first edition of this book.

Power plant construction projects are not performed in isolation. There are many parties involved, which requires a coordinated process for managing safety. First, there must be a process to prequalify contractors. Second, a selection process must be established to ensure that only the right contractors are given the opportunity to bid for work. Third, once contractors have been selected to participate in the project, the prejob planning can begin. The fourth step in the process is for everyone, contractors and owners, to monitor and manage the work-in-progress activities toward achieving the results established for the project. Finally, the fifth step is the postmortem evaluation of the various contractors to determine if they should be allowed to participate in future projects.

EVERYONE SHOULD THINK SAFETY every time a job is started. The safety equipment that is provided must be used. Questions should be asked, and the workers should be encouraged to ask if there is any uncertainty regarding the safe way to perform the work or how to use the safety equipment. Owners, contractors, and workers must make the safety program the best that it can be. An effective safety program is one of the best methods a business can use to stem the profit drain from equipment damage, personal injuries, and associated litigation.

There are various ways to approach the management of job-site safety. In addition to the usual method of using a carrot and stick to reduce accident statistics, there are programs such as the OCIP, in which the owner assumes the risk for managing the site safety program and the contractors do not, then, add the costs of insuring these risks to their contracts. There are programs that focus on behavior, like the behavioral-based safety program that shifts the emphasis away from the number of accidents and toward the number of positive safety behaviors or acts. There are programs that do away with scrutinizing standard statistics like OSHA recordables, lost workday injuries, and so forth. Instead, they only focus on achieving zero accidents; no other measurement has meaning. With the industry safety performance having neared a plateau at many companies, less traditional approaches toward managing safety must be considered, and these are some options.

Owners and contractors must spend time relating the cost avoidance of good safety management to the bottom line of the company. There are many, many ways to manage the safety process of any power plant construction project, but whatever method is used, it should be one that will drive down the potential cost of faulty safety actions. Working safely and generating profits go hand in hand. Anything less is simply a poor method of management.

References

1 U.S. Department of Labor, Bureau of Labor Statistics. *Census of Fatal Occupational Injuries—Hours-Based Rates.* 2010. http://www.bls.gov/iif/oshnotice10.htm

2 Christensen, Theodore A. "Liberty Mutual Loss Prevention." In *ENR Construction Facts*, 2003.

3 Pavey, Rob. "Vogtle Workers Given 8,744 Drug/Alcohol Tests in Past Year." *The Augusta Chronicle*, March 14, 2013.

Managing the Financials 10

Today, it is still all about the money. Managing the construction activities of a power plant project requires a much greater focus on the costs than it did years ago. Many of the plants that are running today are running for one reason only: to generate revenues that will flow to their owners and investors. Just note the number of power plants that are no longer owned by traditional utilities but instead are owned by financial institutions such as banks and corporations formed solely for the purpose of owning and running individual power plants or sites, so that they can provide a return to their shareholders.

Granted, there are also altruistic reasons that some of the more environmentally friendly renewable energy power plants are built. But even then, the investors in these plants get their desired returns through higher rates charged by those selling this power, as well as cost offsets through governmental tax subsidies. Remember what happened to the wind power industry in 2012 when it looked like the U.S. Congress was not going to renew tax credits for building wind farms? Many developers scaled back their build-out efforts since the cost model no longer made these efforts profitable. The result was that fewer plants were ordered until the tax credits were reinstated.

Today, power generation facilities are built and maintained to generate power for sale at prices that will net the investor an acceptable return. This mind-set affects the construction of new power plants and overhaul of existing power plants. If the investor feels that the cost to complete the plant or to rehabilitate it will reduce the return on investment below a precalculated level, the project will not proceed. Therefore, if a project has survived through all of its stages to finally reach the construction phase, it is extremely important that this remaining phase be scrupulously and financially managed as well.

To be able to do this in today's environment requires a special knack for being able to see the future. In addition to understanding man-hour control and how to watch over the costs of supervision and tools, site managers have to be able to predict the outcome of the finances in time to (a) notify their management of impending issues before they become major and (b) take preemptive actions to jolt a poorly trending job out of its doldrums. It is no secret that some of the largest coal-fired plants built in the past few years "surprisingly" exceeded their costs by wide margins. Definitely, some of the causes were out of the *control* of these management teams. But with the proper tools and with proper financial management processes, these cost overruns should not have been surprises.

So how does one prepare? The first step is to become aware of the need for total financial management, rather than simply financial reporting, of the construction process. As with most other stages of the construction process, the stakeholders dictate specific financial requirements that must be met. These range from the usual reporting requirements that provide an indication of how the money has been spent versus how it was planned to be spent to asking for predictions of how the finances will look once the work is complete. The plant stakeholders of today want this information not only periodically but also instantaneously. They want this information so that they will be able to make financial decisions in time to protect or enhance their investments.

These investors are not usually looking for a way to shed their investments; they usually want to maintain or increase their return on investments. This is where good financial management of the construction phase can have a significant impact. By accurately tracking the costs of the work in progress, by actively predicting the cost to complete, and by having these data available instantaneously, owners and investors will have the information they need, when they need it, to make their financial decisions.

The next step is: Don't try to be a magician. It does not work. It will only delay the inevitable. Instead, be a manager. Take a look at previous projects and see what would have made it easier to manage them. Look at projects that were deemed successes and some that were not. Look at projects that were personally managed, and look at some that were managed by others. And most important, do not assume that the information needed to prepare standard company reports will be adequate for also managing the financials of the site work.

A correct estimate and a realistic schedule are required. If the work involves a greenfield site, many details may have changed since the initial estimates were prepared. If the work is a major outage, scope creep may have occurred as the months slipped by since the previous outage. The greenfield project may have incurred changes in schedule due to delays in obtaining permits. The major outage may suddenly see an increase in scope due to a variety of plant availability issues, but without an extension of the schedule. Any of these will impact the original estimate and duration. Therefore, it is extremely important for the site manager to verify the estimate *and* the schedule. For more on verification, see the next section.

Once the estimate and schedule have been verified, a financial status report must be prepared, containing the budget, expenditures, and projection to complete. This may or may not mirror what the home office or other stakeholders require. Almost every job is unique. Therefore, almost every job requires some specificity that others do not. For example, a project in a remote location, such as the one about the 1,000-MW power plant discussed in the earlier chapters of this book, may require special efforts for bringing in small tools and consumables. In such a case, it is often not realized that the cost of special transport for these items can overwhelm the budget allocated for them. So not only does the budget need to account for this, but the financial reporting also needs to address it as well.

Once these fundamentals are understood, the actual work efforts can be addressed. Since the single largest cost component of the site works usually is the cost of the labor, this must be managed in a consistent manner. But it should not be managed by the cost expectations shown in the budget. It needs to be managed through a process of value attained for effort expended. This is often referred to as earned value management (EVM), and it is based on a system of work breakdown structures (WBSs). As will be described in more detail later in this chapter, using EVM and WBSs provides consistency in determining what percentage of a project has been completed at any point in time. It allows for developing the all-important cost to complete. Added to the cost already expended, this cost to complete will then provide a clearer understanding of where to project is headed—good or bad.

Finally, there is one more tool that provides a very powerful indicator of how a construction project may end: a trending view of costs being expended versus value being obtained at any point in time from yesterday to today to tomorrow. It can be thought of as a crystal ball for gauging the chance of success of the project. The financials of a project can be managed by using this trending view of costs along with verified estimates, schedules, and proper budgets and financial reporting.

Validating the Estimate and Schedule

First, validate the estimate. When it was initially prepared, various assumptions were used. These need to be reviewed. For example, most estimators assume that a timely NTP (notice to proceed) will be issued. Since most contracts do not allow for starting work on-site until some type of written notice is provided, a late notice can jam up a project from day one.

Another assumption is that the materials arriving will arrive in good condition. But what happens if some of the materials are not in good condition? Refer to figure 3–4 in chapter 3. Does the estimate take this into account? Does the schedule have time for repairs or replacement? And what about the existing plant, especially if the work is a major pollution control equipment tie-in, or a very busy unit outage in an operating plant? Are the assumed conditions the same as when the initial estimate was prepared?

Also integral to the validation of the estimate is the validation of the schedule. Has it remained the same? Has it been adjusted for seasonal variations in the event of a change in start dates? What about duration? Has the schedule been compressed due to late releases of permits, a delayed NTP, or maybe a change in the outage period due to dispatch issues? There are many reasons a schedule changes, often more so than items affecting the estimate.

The estimate usually includes assumptions about the work process and progress. For example, if there will be heavy concrete work, was the initial estimate based on using an on-site batch plant? But is this still the case, or has something changed forcing the concrete to be brought in by truck? That would certainly impact the cost for this portion of the work scope as well as potentially impacting the schedule.

What about welding? Most estimators assume a specific number of man-hours per weld, or equivalent weld. As shown in figure 3–5 in chapter

3, one way of developing the man-hours required for welding tube ends is to apply an estimating factor of man-hours per equivalent weld. (More on equivalent welds is in chapter 11.) However, this factor is usually a result of historic data, and the welders on the current project may not be able to achieve this productivity. In that case, the estimate will not be reflective of what will happen on the current project, and another approach may need to be investigated, such as using automated welding.

In general, if any of the original assumptions have changed since the estimate was prepared, the new conditions must be considered when preparing the financial management tools. This may require a revaluation of the estimate and related budget, which is essentially a risk analysis, and the budget may need to be increased or decreased. The same applies to the schedule. It could be impacted by a change in any of the assumptions, and a change in schedule almost always has a corresponding impact on the budget.

Doing all of this may be as simple as reviewing the original assumption list, providing that old-fashioned gut feel, and saying, "yes, that's exactly what will happen." Or, it may require a more complicated review, one that requires a Monte Carlo type of analysis of some of the risk assumptions. For example, for weather-related assumptions, a search of weather patterns over the last 50 years may show that in some years there would have been nine lost days, and in other years, none. Applying a Monte Carlo simulation would enable the site manager to predict, within a prescribed factor of accuracy, the actual number of days of delay that might be encountered. Additionally, upper management may relate, understand, and be more comfortable with the probability of finishing on a certain date after this type of analysis.

Finally, after reviewing and assessing the assumptions that were made when the estimate and schedule were initially prepared, and after analyzing site processes and productivity, there is still one more important validation tool—benchmarking. Does the estimate reflect historical patterns from previous projects? Are adjustments necessary due to different locations, different clients, different labor sources, and so forth? What about industry standards. Does the estimate reflect what others in similar industries report?

Preparing the Financial Status Report

Structuring the report

The next step in the process of preparing for managing the finances of the project is to prepare the financial status report. This is a report designed to compare the ongoing expenditures to the baseline budget, then add in the estimated cost to complete, and use these data to project the cost at completion. This report should include categories similar to those used when the work was estimated, such as labor, supervision, tools, equipment, small tools, and consumables. The spreadsheet report shown in figure 10–1, taken from an actual boiler outage, is one way to prepare such a report.

Note that the first column of the spreadsheet is labeled *Description*. This column is designed to capture the costs in logical categories, as discussed above. However, it is important to realize that these categories are for summary purposes as part of high-level reporting. Somewhere there must still be a much more detailed breakdown of the work and its reporting categories. For example, the work should be divided into as many WBSs as will be practical to report against. If the job entails the replacement of superheater elements, there should be a subcategory for removal, another for milling and beveling the header nipples, another for installing the new elements, and also one for welding out the replacement elements. These subcategories are also important in measuring and managing productivity. The man-hours expended in these subcategories should be rolled up to a higher level WBS, which in turn should be "rolled up" to a summary category such as the *Field Labor* category in figure 10–1.

Further examining the example in figure 10–1, the *Supervision* and *Field Accounting* categories includes the payroll costs of the supervisory and administrative personnel as well as any travel and living expenses associated with their work. The *Small Tools and Consumables* category generally includes items such as hand tools, cutters, reamers, and other tools that have a short shelf life, are not expected to be reused once the job is complete, and/or need to be replaced at least once a year—in other words, items that the accounting department will want to expense rather than capitalize. Similarly, items called consumables, such as rags, oils, greases, grinding wheels, and any other item that is not reusable, also fall into this category. When these items have to be shipped from elsewhere, the cost of shipping them should be tracked. As mentioned earlier, on remote sites, these shipping costs can become substantial, sometimes rivaling the cost of the small tools and consumables themselves. In these cases, it is prudent management to plan ahead and prepare a few large orders that can be shipped by surface freight instead of falling into the trap of making multiple emergency orders that require air shipments.

The next category, *Tools and Equipment*, is generally used to report items that are expected to be reused on future jobs and last more than one year—for accounting purposes these items will be capitalized and depreciated, not expensed at the time of purchase. These items include chain hoists, grinders, and drills. Equipment includes welding machines; milling equipment; and large lifting devices such as cranes, hoists, trolleys, and jacks. The freight costs for this type of construction apparatus can be significant. For example, in addition to the costs for transporting a large crane to and from the site, there may be costs for temporary preparation of roadbeds and bridges, removal of overhead obstructions, and possibly special police escorts.

Continuing with figure 10–1, the items in the *Materials* category are usually purchased at the job site and fall into two categories: those that form part of the final product and those that do not. Often, it is important to differentiate between the two due to tax issues; the material forming part of the final product may be eligible for tax exemptions if the final product is tax exempt, whereas the temporary construction materials may not be. Weld rod would be a classic example of a material forming part of the final product. Scaffold boards and fit-up bolts for structural steel would be considered temporary construction materials. Again, freight may be significant if the job is at a remote location.

Often, subcontractor costs are managed as a package. In other words, many contractors and owners only look at the bottom line of the costs from their subcontractors, and this is OK when there are no problems. This is what has been portrayed in the *Subcontractor* category of figure 10–1. However, when the subcontracted portion of the job becomes a significant percentage of the overall project, these costs should be detailed and reviewed with the subcontractors as if the subcontract did not exist. Therefore, this category sometimes needs to be expanded. The expansion would allow subcontractor costs to be compared to their budget and in the event of shortfalls, discussions could be held immediately to develop a plan of recovery.

Financial Status Report

Description	Budgeted	Expended to Date	Percent Expended	Projected at Completion	Projected Variance	Variance Percent
Field Labor Wages	4,006,000	4,009,000	100.07%	4,075,000	(69,000)	
Bonuses/Vacation	30,000	0	0.00%	10,000	20,000	
Total Field Labor ($)	**4,036,000**	**4,009,000**	**99.33%**	**4,085,000**	**(49,000)**	**-1%**
Total Field Labor MH	67,255	64,063	95.25%	65,000	2,255	
Avg. Wages PerLabor MH	60.01	62.58	104.28%	62.85	(2.84)	
Supervision Wages	431,885	288,155	66.72%	338,000	93,885	
Supervision Expenses	106,674	120,164	112.65%	116,674	(10,000)	
Total Supervision ($)	**538,559**	**408,319**	**75.82%**	**454,674**	**83,885**	**16%**
Total Supervision MH	5,081	4,367	85.95%	5,081	0	
Avg. Wages Per Supervision MH	85.00	65.98	77.63%	66.52	18.48	
Field Accountant Wages	52,524	41,210	78.46%	78,016	(25,492)	
Field Accountant Expenses	25,842	23,557	91.16%	25,842	0	
Total Field Accountant ($)	**78,366**	**64,766**	**82.65%**	**103,858**	**(25,492)**	**-33%**
Total Field Accountant MH	1,094	1,496	136.72%	1,625	(531)	
Avg. Wages Per Accountant MH	48.01	27.55	57.39%	48.01	0.00	
Small Tools & Consumables	75,000	185,500	247.33%	175,000	(100,000)	
Office Costs and Supplies	40,000	21,677	54.19%	23,000	17,000	
Freight on Sm. Tools & Cons.	11,000	8,536	77.60%	9,000	2,000	
Total Sm. Tools & Consumables	**126,000**	**215,712**	**171.20%**	**207,000**	**(81,000)**	**-64%**
Percent of Field Labor Wages	3.15%	5.38%		5.08%	(0.02)	
Lifting Equipment (cranes, hoists, trolleys,etc.)	30,000	43,441	144.80%	50,000	(20,000)	
Welding Machines	15,000	24,861	165.74%	24,861	(9,861)	
Milling Tools	10,000	4,941	49.41%	4,941	5,059	
Others	74,000	66,789	90.26%	68,500	5,500	
Freight on Tools & Equipment	10,000	11,446	114.46%	12,500	(2,500)	
Total Tools & Equipment	**139,000**	**151,478**	**108.98%**	**160,802**	**(21,802)**	**-16%**
Cost Per Field Labor MH	2.07	2.36		2.47	(0.41)	
Weld Rod	10,000	10,823	108.23%	12,000	(2,000)	
Contract Material	15,000	15,051	100.34%	15,000	0	
Temp. Construction Material	17,000	2,238	13.16%	10,000	7,000	
Freight on Materials	9,500	1,500	15.79%	1,500	8,000	
Total Material	**51,500**	**29,612**	**57.50%**	**38,500**	**13,000**	**25%**
Insulation / Lagging / Abatement	211,000	198,000	93.84%	211,000	0	
NDE (including UT)	35,000	25,000	71.43%	15,000	20,000	
Stress Relieving	58,349	74,000	126.82%	74,000	(15,651)	
Other Supervision	8,500	0	0.00%	0	8,500	
Total Sub-Contractors	**312,849**	**297,000**	**94.93%**	**300,000**	**12,849**	**4%**
Builders Risk Insurance	5,500	0	0.00%	3,820	1,680	
Insurance	**5,500**	**0**	**0.00%**	**3,820**	**1,680**	**31%**
Contingency	0					
Profit Margin	325,000					
Contingency & Margin	**325,000**					
Company Overhead	775,000					
Total Overhead	**775,000**					

					Net Profit	
Totals:	**6,387,774**	**5,175,887**	**97.88%**	**5,353,654**	**259,120**	**4.06%**

Report Period Information:
1.) Number of Workers Current Week:
Craft Workers: 66
Supervisors: 5
Total Force: 71
2.) Number of Shifts Currently Working: 2
3.) Hours Per Shift Currently Working: 10
4.) Days Per Week Working: 7
5.) Physical Percent Complete: 98%

Current Week Information:
1.) Number of Workers Current Week:
Craft Workers : 25
Supervisors: 4
Total Force: 29
2.) Number of Shifts Currently Working: 1
3.) Hours Per Shift Currently Working: 8
4.) Days Per Week Working: 5
5.) Physical Percent Complete: 99%

Fig. 10–1. A financial status report (Courtesy of Construction Business Associates LLC)

Some contractor managements prefer to conceal contingency, profit margins, and overhead costs from the field staff. Certainly, most do not want to expose this information to their competitors. Although the example in figure 10–1 includes this information, it is not crucial for the management of the site activities. It does, however, provide a picture of how the site work will impact the overall financial results to those responsible for the financial well-being of the project.

Finally, to offer a complete picture of the costs affecting the job, a synopsis of the on-site personnel is frequently helpful. At the bottom of figure 10–1, two time periods are shown. One reflects the week the costs were recorded, and one that reflects the current point in time. The difference is generally due to the time lag between when the costs are captured and when they are recorded, often a week. The same is true for the percentage of completion, so it is also shown twice.

Once the categories for the financial reporting have been established, the job estimate can be subdivided and entered into the next column, here labeled *Budgeted*. Sometimes, it will be difficult to take the estimate as prepared by the estimator, or as sold by the salesperson, and apply it to the categories previously described. However, it is important to find a way to do this so the work can be managed properly and feedback can be provided for estimating the next job. At times, it may require that the site manager and the estimator sit down to redistribute the numbers in the estimate in accordance with the categories required for managing the job.

As this spreadsheet reflects, having determined which categories to use for reporting, and having distributed the budgeted numbers into these categories, the rest of the spreadsheet is easily completed. Columns are established for (a) the amount expended, (b) amount expended as a percentage of the budgeted amount, (c) the amount projected at completion, and (d) how far the amount projected deviates from the budget.

Gathering the data

The next step in preparing to report the financial status of the job is to set up a system of collecting costs and entering them into their categories on the spreadsheet. The first information required is the amount already expended. This can often be problematic because accuracy is important. The timing of data availability often becomes very frustrating. Although the actual man-hours and resultant payroll costs are never more than a few days old—due to the need to prepare the workers' paychecks in a timely fashion—the supervisors are often weeks behind in submitting their time sheets and expense reports. The vendors from whom materials are purchased and from whom equipment is rented are sometimes months behind. Most site staffs do not have the luxury of chasing after these costs while the job is ongoing so the costs recorded in the *Expended to Date* column are generally only those costs captured to date by the accounting system. This column is essentially a representation of the flow of the cash-out-to-date.

The next column, *Percent Expended*, only represents what is in the accounting system, not necessarily what has been spent. The following column, *Projected at Completion*, is the heart of the report. It actually comprises two subcolumns (not shown), the first of which approximates the difference between the expended numbers as reported in the accounting system and the debt still owed that has not yet been entered. This first subcolumn is used to keep track of the site staff and commitments made to vendors. For the site staff, someone should maintain a log of all supervisory personnel charging into the job. This log would then be used to approximate the supervisory costs already incurred, although not yet reported and claimed.

For the vendors, which also include subcontractor billings owed but not yet received and/or processed, a log of commitments made to date is a useful tool that can be used to determine how much still has to be paid for work already performed. This forms the basis of the first (hidden) subcolumn.

The second (hidden) subcolumn is made up of the projected costs from the present day to the end of the job. When added to the first, it forms the total projected costs to complete the work. Calculating the percent of variance will then quickly show where problems may be lurking. Looking at figure 10–1, after inputting the costs to date, adding those costs owed but not yet collected by the accounting system, and then adding the costs anticipated to complete the project, the message is clear. One can see that there is a 1% overrun projected for the labor budget and significant overruns of the accounting and all tools and equipment costs. The supervision budget will have some money left over, as will the materials and the subcontracted budgets. All in all, this job, which is essentially complete at this point, will be profitable for the contractor, but not as profitable as planned; due to a 1.2% expected cost overrun, the profitability of the job gets hit with a 20% reduction—a tough wake-up call now that net profit is projected at $259,120 versus the budgeted $325,000.

Comparing the Actual with the Expected and Predicting the Results

Although the financial status report, as shown in figure 10–1, is an important tool for managing the finances of the project, it is a static tool showing the financial status of the job at a single point in time. To be able to predict the future, more is required. Comparisons must be made between what was planned and what has happened. Then, predictions can be formulated by viewing the trends the data portray; the cost to complete can be calculated; and if necessary, actions can be taken to impact those trends. This is the crux of the process of managing the job through its finances— projecting the trends and taking pre-emptive action when required.

Earned value management (EVM)

But before we can talk about costs and trends, we need to understand their underlying WBSs and the EVM process of determining work accomplished versus effort expended.

Work breakdown structures. WBSs are established for ease in reporting costs as the project progresses. However, they also have two other important purposes: (1) as a tool for measuring the progress of the work and (2) as a tool for projecting the outcome of the job, both in terms of schedule and in terms of cost.

Because of their use as tools for these last two services (measuring progress and projecting outcome) the WBSs must be assigned realistic values of importance relative to their overall impact on the job. For simplicity, let's look at figure 10–2, a job with four major activities, or WBSs. The first activity shown is the turbine reblading. This was estimated to require 15,000 man-hours to complete. The next activity, condenser retubing and waterbox repairs, was estimated to require 83,000 man-hours to complete. The third activity, pressure part replacement, was estimated to require 82,000 man-hours. The final activity or WBS, other, was estimated to require 150,000 man-hours. As can be seen, the weighted value of each WBS has been calculated as a percent of the total man-hours of the project. This serves as a sanity check to ensure that each estimated WBS is valued in proportion to its expected impact on the overall work to be undertaken. As an example, if the turbine work were to be arbitrarily assigned a weighted value of 50% of the project, the above would show that as being unrealistic.

Weighted Values		
WBS	Man-hours Budgeted	Weighted Value
1. Turbine Reblading & Repairs	15,000	5%
2. Condenser Repairs	83,000	25%
3. Pressure Parts Replacement	82,000	25%
4. Other	150,000	45%
Total	330,000	100%

Budgeted WBS Man-hours ÷ Total Job Man-hours = Weighted Value

Fig. 10–2. Weighted values

Earned value. Once the WBSs are weighted and accepted, they can then be used for determining progress and predicting outcome. As an example, when the WBS representing the condenser work shows that the work is at its halfway point, the time for its completion can be considered half spent, as well as its costs. Looking at figure 10–3, one can see the steps required to determine this.

Earned Value Calculation – Step A			
WBS	Man-hours Budgeted	Percent Complete	Man-hours Earned
1. Turbine Reblading & Repairs	15,000	35%	5,250
2. Condenser Repairs	83,000	50%	41,500
3. Pressure Parts Replacement	82,000	40%	32,800
4. Other	150,000	60%	90,000
Total	330,000		169,550

Budgeted WBS Man-hours × % Complete = Man-hours Earned

Earned Value Calculation – Step B			
WBS	Man-hours Earned	Earned Value	Man-hours Expended
1. Turbine Reblading & Repairs	5,250	2%	6,000
2. Condenser Repairs	41,500	13%	38,000
3. Pressure Parts Replacement	32,800	10%	37,500
4. Other	90,000	27%	92,500
Total	169,550	51%	174,000

Man-hours Earned ÷ Budgeted WBS Man-hours = Earned Value

Fig. 10–3. Earned values

In step A in figure 10–3, the actual percent complete must be determined for each WBS. This is done at the lowest level possible and then rolled up to the highest level of WBSs. For example, the condenser repair work in this example included retubing and waterbox repairs. For the sake of argument, let's say that the retubing work and the waterbox work each made up half of the condenser's 83,000 man-hours. If the condenser had 25,000 tubes, and 6,250 had been removed and replaced, this would suggest that 25% of that planned work was complete. Say that at the same point in time, 75% of the waterbox repairs were complete, then an experienced construction person might conclude that 50% of the total condenser work scope was now complete. This is an example of rolling up smaller WBS categories into the larger condenser WBS.

Next, the man-hours earned are calculated from the percent complete determination of each WBS. As shown, this is done by multiplying the budgeted man-hours by percent complete. Then, the earned value of each WBS is calculated by dividing the man-hours earned by the budgeted man-hours, as shown in step B of figure 10–3. It is very important to note that this value, the man-hours earned, *does not* necessarily equal the man-hours expended! Note that in the last column of figure 10–3, the shaded column, the actual man-hours expended are recorded. And note that not in any one instance do they equal the man-hours earned. They are independent of each other—this cannot be overstressed.

Again, let's refer to the example of the condenser repairs. This work was budgeted at 83,000 man-hours. When it is 50% complete, it will have *earned* 41,500 man-hours, even though only 38,000 man-hours were *spent* to reach this point. And regardless of how many man-hours it takes to complete the condenser repair tasks, upon completion, they will have *earned* only 83,000 man-hours. This then also reflects 13% of the total cost of the project to date, because half of its 25% impact on the job has been spent. (Numbers are rounded to avoid decimals.)

Now let's look at the total job. At the point in time indicated in the example, when the condenser work is 50% complete, the turbine work is 35%, and so on, the overall completion status of the outage is 51%. This was determined by calculating the earned value of each WBS at this point in time—by multiplying its physical percent complete by the weighted value of the WBS—and then adding them together to arrive at 51%. However, note that 53% of the budget has been spent. And how was this determined?

It is the sum of all the *actual* man-hours that were spent divided by the total *budgeted* man-hours: 174,000 man-hours divided by 330,000 man-hours. So what does this mean? It says that for 53% of the man-hours expended to date, only 51% of the work has been *earned* (the earned value).

Cost to complete

As shown in figure 10–4, these results suggest that if no action is taken and the remainder of the job continues in a similar fashion, at the completion of the work, the budget will be exceeded by 3%. To determine this, the man-hours expended to date were divided by the attained percent complete to obtain the projected cost at completion for each individual WBS. These projections were then added to arrive at the total projected man-hours for the project. This projected 3% overrun of the outage budget translates into an extra 11,060 man-hours that will be required to complete the work (341,060 projected man-hours minus the originally budgeted 330,000).

	Projected at Completion			
WBS	Man-hours Expended	Percent Complete	Man-hours at Completion	Budget at Completion
1. Turbine Reblading & Repairs	6,000	35%	17,143	114%
2. Condenser Repairs	38,000	50%	76,000	92%
3. Pressure Parts Replacement	37,500	40%	93,750	114%
4. Other	92,500	60%	154,167	103%
Total	174,000		341,060	103%
Man-hours Expended ÷ % Complete = Man-hours at Completion (if nothing changes)				

Fig. 10–4. Projected at completion

Once this analysis has been completed, a decision needs to be made as to whether or not these new numbers are acceptable. Assuming that they are, then the financial report, as shown in figure 10–1, should be updated and reissued. This report would then show the new job projections.

However, if these new numbers are not acceptable, other means of completing the work should be explored. As these other ways are reviewed, the remaining work should be re-estimated to see if the cost to complete can be reduced. In this case, not only should the man-hours be re-estimated, but also all of the other categories that make up the financial report should be re-estimated. Unless a job is going extremely well, this updating process needs to be performed frequently to find potential problems before they become surprises.

All of this is a straightforward way of seeing what might happen if nothing is done to change the course of the project. It affords the site management staff opportunities to take corrective actions in time to effect meaningful change. However, other important indicators should be used to complement the above.

Trending

Figures 10–5 and 10–6 each contain two graphs showing the trending of a job and how to use these trends to help manage it. Although based on real-life cases, the graphs have been modified so that each follows the same baseline data. They depict the same project, but with four different possibilities. Each is based on a 14-week outage with a budget of 330,000 man-hours; total job costs are not used for several reasons. The first reason, as previously pointed out, is that the actual cost data often trails by weeks or months. The second reason is that most construction budgets are actually based on the man-hours required to do the work. Therefore, using man-hours as a cost indicator is generally a good indicator of the total job costs. Third, since worker payrolls must be prepared at least weekly, the man-hours are usually the first data that are available.

These graphs each depict four pieces of information that when viewed together provide a strong indication of the direction of the job at an early stage, in time to take preemptive action if necessary. The left-hand axis is a measurement of the number of man-hours of the work, and the right-hand axis denotes the percent complete. The intent of these graphs is to compare the man-hours expended and the actual percent of completion achieved for these man-hours with the planned percent complete curve. By making comparisons among the man-hours expended, the percent achieved for those man-hours, the man-hours budgeted, and the percent planned, one can readily see if the job is heading for trouble or success.

The first graph in figure 10–5, Progress Graph 1, shows a well-planned and well-executed job. The man-hour expenditure curve remained below the planned percentage of completion curve and below the actual percentage of completion curve. The actual percentage of completion curve remained above the planned percent complete curve. This is referred to as a positive variance, or Δ (delta). In contrast, the second graph, Progress Graph 2, shows a job with a negative Δ; it is a job in trouble.

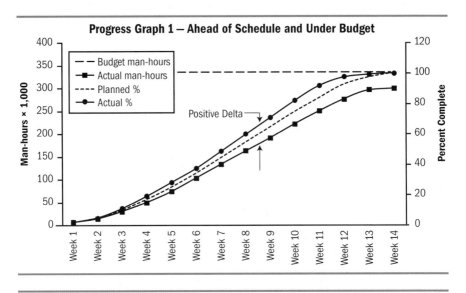

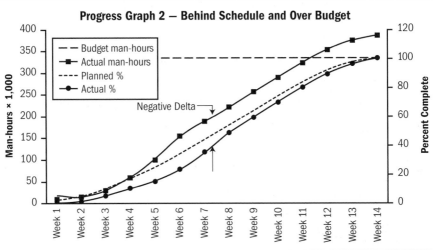

Fig. 10–5. Progress graphs (Courtesy of Construction Business Associates LLC)

The man-hour expenditure curve in this second progress graph is above the planned percentage of completion curve and above the actual percentage of completion curve. The actual percentage of completion curve is below the planned percent complete curve. This is referred to as a negative variance, or Δ, and basically says that more man-hours (or money) are being spent than progress being made. It is usually difficult to get out of this situation, but not impossible if caught early enough, as is demonstrated by the graphs in figure 10–6.

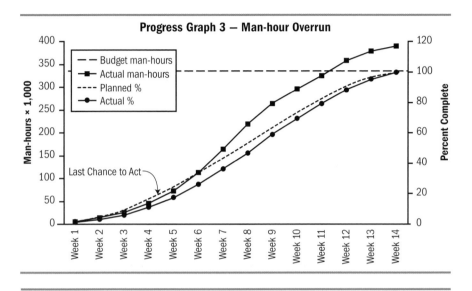

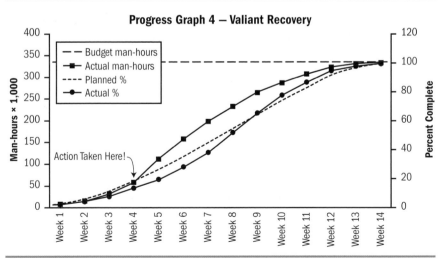

Fig. 10–6. Progress graphs (Courtesy of Construction Business Associates LLC)

Progress Graph 3 in figure 10–6 similar to the previous graph, shows a job with a negative Δ. But note that until about the sixth week, both the actual man-hours being spent and the percentage actually being gained are below the planned percent curve. This suggests that there was still a chance to complete the work on time and without going over budget, but action needed to be taken back in the fourth week. Following the timeline until the end of the job, however, shows that the action taken was seemingly just a gradual, possibly daily, ramp-up of personnel that did not jolt the job out of its slump. This job did complete on schedule, but at a cost of almost 20% above budgeted man-hours, and *this was predictable from the very beginning!*

In contrast, Progress Graph 4 also shows a job with a negative Δ, or in trouble just like the previous one, but with the difference that drastic action was taken. Note that in the fourth week, the personnel were dramatically ramped up. This created an increase in the rate of completion, which met with the planned percent complete curve by the ninth week. Thereafter, the personnel were gradually reduced, and the job completed on time *and within the budgeted man-hours.*

Using these methods of graphing the man-hours being expended on a project, and comparing them with both the percentage of completion expected and the percentage of completion attained, one has the opportunity to make decisions in time to impact the outcome of the job. The key, however, is to start this process at the very beginning, closely monitor the trends, and make changes early, in time to still have an impact on the job.

Looking once more at the last graph in figure 10–5 and both graphs in figure 10–6, by the fourth week their percent complete curves were trending further and further away from their planned percent complete curves. Their man-hour expenditure curves were closely tracking their percent complete curves, suggesting that the man-hours being expended for the progress being attained were essentially at the right level. But the actions taken on the job represented by the fourth graph (the valiant recovery) showed that an early dramatic ramp-up in man-hours can pull up the completion percentage and even allow time to ramp back down to stay on budget. That's what was done, and it worked.

Summary

Again, it really is all about the money. The site managers of today are obligated to tell their management and investors what is happening with the money. Not only what happened with the money just spent, but also what is expected to happen with the rest of it. They must understand this responsibility. They must know why their stakeholders are interested, and they must have the tools and support to provide the information their stakeholders and investors will want, when they want it.

Site managers have to understand what information is needed to manage risks. But to do so, they also have to understand what risks could be incurred. Managers do not have a crystal ball that can predict the future, but most do have the background experience that offers some insight into what might happen; if they do not, they usually have access to someone who does. Properly formulated, a brainstorming session involving past and present project participants can lead to a wealth of ideas on what to track, how to use this information, and what to focus on. Reflecting on past experiences can provide very useful guidelines for managing the future.

Once a clear picture has been developed for what is needed, the project expenditure categories can be formulated. This will lead to a reporting format, namely, the financial report, that is sensible for the project, is a tool for predicting the future, and will be a source of invaluable data when the next project is being bid.

But the financial report is just that—only a report, a repository of historical data. The job-site managers of today cannot stop there. They must extrapolate these data to predict the future, or at least to give the investors some idea of what might happen if the present course of action continues. They must trend the historical data, preferably with easily read visual charts that can project the job to completion. They must do this early in the project and regularly during the project. Doing this preserves the opportunity to take action that can impact the job's outcome, in time to realize the results that were wanted, thereby avoiding unpleasant surprises. Without this, that opportunity may be gone—there is a maxim that once a project has reached the 20% completion point, the opportunity for change is almost gone.

There are tools that can be used to extrapolate the results from a current point in time to project the future. One such tool is the graph that compares the plan with the actual. Projecting the actual into the future shows whether things will go right or wrong and offers the opportunity to make a change in time to impact the job. This is called "trending the job," and it is very powerful.

Another necessary tool is the cost-to-complete model. This tool can suggest where the cost at completion may be if no change is made. It will not suggest when to take action, but it will suggest what may happen if action is not taken (which may be OK). However, it is very important to capture the costs by categories that lend themselves to the actual activities of the job—the WBSs. At the same time, it is necessary to communicate the timeline basis of the data and the timeline of today; data skewed due to timing of results must be used accordingly.

All of these tools are theoretically good, but are they valid? This is an important question; validation of underlying assumptions must be part of the plan. Is the estimate right? Are its assumptions still valid? What are the chances that these assumptions have changed? Has the estimate been benchmarked—both in-house and in the industry?

Then there are those WBSs. What are they, how were they developed, and how will they be used? First, they serve as a structure for collecting the costs. Second, they become a tool for measuring progress and projecting outcome, and this is possible because of their individual impact on the job, their weighted value.

Finally, a step toward productivity must be made. Although the subject of productivity is reserved for chapter 11, it requires linkage to earned value, which is described herein. Earned value is the measurement of results against efforts. It is not an earning for effort expended, but an earning for results gained. Regardless of the effort expended, 50% gained is just 50% gained!

Managing the Site Activities and Cost Control 11

(with content from Mark Bridgers)

anaging the job-site finances, the subject of chapter 10, does not, by itself, guarantee a successful outcome. It provides early warnings of problems, which can be used to make changes to alter the future. But what changes should be made, and what can be changed? There are a host of areas where changes can be made, such as the following:

- Labor
- Schedule
- Tools and major equipment
- Consumables
- Materials
- Site services

Making changes to the way any of these areas are managed will impact the job. Changing the way labor productivity and overtime are managed can dramatically alter both the schedule and the cost of the project. A change in the schedule, either compressing or extending it, will affect the final costs. Making a change in the way the tools and equipment are being used or managed can influence the end result. The same is true for the other areas. Since all of the areas listed above are integrated to form the total project, making changes in one will cause an effect in the others.

The purpose of this chapter is to delve into these areas and explore how they should be managed and how to make changes, if necessary, to realize a net change to the bottom line. In other words, this chapter is about making cost-effective changes. Tools such as productivity management and overtime will be discussed. The importance of proper schedule management will be demonstrated. How to manage tools and major equipment, consumables, materials, and the site services will be addressed. How to stay on top of it all, as well as knowing when to look deeper, will be shown with a dashboard example. E-commerce and the Internet will be addressed along with bar coding and a few of the modern technology tools.

But before exploring how to manage the various segments of the job, a determination must be made of what is important and what is not. Not all of the items listed above are important for all of the projects, all of the time. Not all jobs will require large, heavy equipment. Not all jobs will be involved in purchasing materials. The importance of some of the segments will change as the project progresses. It is important to know where to focus one's attention, and when, to maximize the return for that effort.

Labor

To assist in this determination, an evaluation of each area as it relates to the overall cost potential of the project is important. Labor is usually the single largest component of any construction project. For a grassroots new power plant project, the labor costs can range anywhere from 35% to 65% of the construction costs, depending on where the job is located, where the labor comes from, and what kind of job it is. In other words, for a combined cycle job, the labor component would be on the lower end of the scale, whereas a new coal-fired project would be more labor intensive. For a major outage, again, location and job scope will be a large determinant of the labor proportion of the job costs, but generally, the labor component will play an even more significant role—which is due primarily to the removal/demolition component of the work, which is not present in new construction projects.

But regardless of the percentage of the job costs attributable to the labor segment, because it is one of the largest factors, managing it is the single most important facet of the project. When the labor is not properly managed, the remaining segments of the project are strongly impacted. Supervision may need to be reinforced; the schedule might need reworking; additional tools and possibly longer durations of costly equipment may be required; the use of consumables may be increased; and safety and quality may suffer as well. All in all, when labor is not effectively managed, the job suffers.

Productivity

Getting a handle on managing labor requires an understanding of what is expected versus what is being achieved, in other words, labor productivity. Labor productivity is a very complex subject. Many books have been written about it. Norms are constantly changing, which keeps the subject open. Here, the purpose is to describe some techniques for managing productivity.

We know that at the site level, there are a host of challenges that constrain the productive efforts of field forces, trades, supervision, and site or construction management. Resource availability, aging workforce, remote sites, and coordination complexity are just a short list of these challenges. Frequently, site and construction management, when pressed for ideas on faster, more efficient, safer, and higher-quality performance, will respond, "we are doing all we can." But are they?

Let's look at a training exercise that can be used to demonstrate the answer to the above question. Here we use a simplistic model with 61 multicolored interconnecting parts that when constructed correctly, will produce a small walking bridge (fig. 11–1). With this simple model, many of the site construction challenges that a construction manager and the construction team will face can be replicated. A host of challenges and self-imposed constraints will become apparent immediately:

1. Time constraints
2. Pressure to perform (verbal exhortations, peer pressure, etc.)
3. Incomplete design documents
4. Incomplete construction drawings
5. Materials storage constraints
6. Small tools utilization
7. Site size constraints
8. Crew size implications
9. Prefabrication opportunities
10. Effectiveness of supervisor or crew leader
11. Preplanning
12. Information access

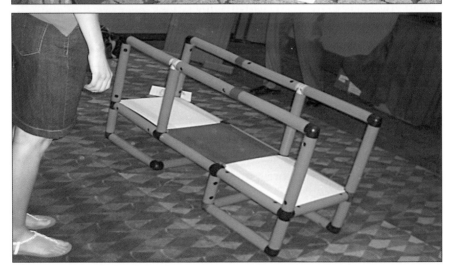

Fig. 11-1. Bridge-building exercise

The training exercise goes as follows: The participants are divided into teams of five or six members each. A folding table is set up against one wall of the training facility. The 61 bridge parts are placed on the table— see the uppermost photo in figure 11–1. The instructor then provides a black-and-white, three-dimensional drawing of the completed bridge and explains that the pieces are to be put together as shown in the drawing. The instructor explains that for assembly, the pieces snap together, but for disassembly, a special tool is required. Then the team is given a fixed time of perhaps six minutes to build the bridge. Only one team works at a time—see the photo in the middle of figure 11–1. The other teams silently watch (and take notes?). When the time is up, the exercise stops, and a count is made of the correctly assembled pieces—seldom does this equal 61. So a score is assigned as a ratio of the number of those pieces to 61.

Next, a brief group discussion of the whole class is held to talk about what the first team could have done differently in order to obtain a better score—or higher productivity. Then the bridge is disassembled, the parts are piled up on the table again, and the second team goes to work. At the end of the same time of six minutes, the second team's correctly assembled pieces are counted and a score is determined. Usually this is still not a perfect 100%, but it is usually higher than the first team's score.

This continues until only one team is left. Then, the instructor asks the group some questions, such as: Did anyone think to ask for more disassembly tools? Did anyone think to just move the table away from the wall, thereby providing access to all four sides of the work platform? Did anyone think to reduce the number of active team members, because maybe the crew was too large for the task? There are many more questions that are asked, but it would be unfair to future participants of this exercise if all were revealed here.

Finally, the last team is asked to go to work armed with much of the information the others did not (know to) ask for. When this last team completes the bridge, it is usually under the time limit, and if they did not complete it, they definitely connected more pieces than any of the other teams (see the bottommost photo in fig. 11–1). A class discussion then ensues, and frequently a major take-away is that we are creatures of habit. We do not readily form and perform as teams, and we seldom ask others for input.

As we saw, the first team just started working. Usually, they ask no questions. Frequently, they have inefficient individual work assignments. They often make incorrect assemblies, wasting time by needing to disassemble and reassemble pieces, and they seldom complete the bridge. On the other hand, the second team, which has been afforded the benefit of some extra information from their own observations of the first team's work and from the class discussions, does a better job. They usually come much closer to completion, but again they seldom complete the bridge either. Finally, the last team, which has been privy to all of the previous teams' trials and errors plus the instructor's additional hints, often completes the exercise in the prescribed time and always correctly assembles more pieces than its predecessors. Obviously, this exercise can be expanded from productivity training into full-blown team building. But that is the subject of other books.

The increase in the number of parts completed by each successive team in the above exercise can be compared to recoverable lost time or waste reduction. In some variations of this exercise, the number of parts requiring disassembly and reassembly are recorded by others. This is done to measure rework, which obviously has an impact on the timely completion of the bridge and figures into productivity.

Based on the numerous times this exercise has been run by Mark Bridgers of the Continuum Advisory Group (and coauthor of this chapter), several important statistics have surfaced. The results from the first team to the last team usually improve by over 150%. Rework occurrences are frequently reduced on the order of 85%, down to less than 10% of the work effort. In other words, lost time is recovered by very significant percentages.

So how can this productivity exercise be applied to the real power plant construction world? First and foremost, the value of preplanning comes to mind. Identify the possible obstacles that will be encountered. Assign specific tasks to specific individuals. Make sure everyone understands their roles and expectations. Then, survey the work area and determine if more is needed—maybe some preassembly work. Also, assess the available tools and equipment and supplement them, if necessary. Whether building a small model bridge or a major combined cycle power plant, the basics of productivity skills remain the same.

Now, let's look at some actual job-site specifics. Here is a listing of some of the categories of work that can be readily measured:

- Electrical
  - Conduit installation, by the foot
  - Wire and cable installation, by the foot
  - Terminations, by the number
- Mechanical
  - Piping installation (large bore, small bore), by the foot
  - Welds completed (by the weld, by the equivalent weld)
- Insulation
  - Flat surfaces, by the square foot
  - Piping (large bore, small bore), by the linear foot

These categories should be tied directly to the work breakdown structures (WBSs) established earlier, during the design of the financial control system. Depending on the work scope, not all categories will be applicable; at times, additional categories will be required.

In addition to selecting the categories, each category should be subdivided into discrete areas of the work, which may or may not be identical to the WBSs. For example, a WBS may be installing cable from the various pieces of equipment to a new control room. But since this work will entail many different pieces of equipment, in many different locations, each major run of cable and wiring should be measured separately because the degree of difficulty for each may be significantly different.

The same applies to welding. The productivity of welding waterwall panels in a boiler with a convenient scaffold in place is very different from that encountered when welding replacement sections of dissimilar metal welds in the superheaters while the welder is lying on his or her back. Access is a major factor, as is the type of welding required, especially if preheating is required.

With the categories of work to be measured decided, expected production can be established. This can be obtained from industry standard references, local labor organizations, or the records of jobs previously performed. These expected production rates then become the norms.

Next, daily measurement of each activity should be made and plotted, similar to figure 11–2. In the first chart of this figure, the welding progress chart, the white bars show the planned or required welding production necessary to complete this activity by the end date. However, by comparing the white bars with the dark bars, it can be readily seen that the initial welding was less than what was established, or put another way, the production or progress for this phase of the work started out lower than required. As the job progressed, the production increased, exceeding the daily requirement for approximately half of the duration, after which it was reduced to enable completion per the original plan.

There are also two progress curves plotted on the production graph in figure 11–2: the number of cumulative welds required to meet schedule and the cumulative welds achieved. When this work first started, it was behind schedule, as shown by these two lines, and it did not reach the required level of production until 11 days later. Then, it actually exceeded the required production for the next 12 days and finally stayed almost exactly on target for the last seven days.

Why not just divide the number of welds by the number of available days and have a level production plan? In this case, 300 welds divided by 30 days arrives at an even production level of 10 welds per day. This might be possible in many instances where no work is required beforehand. In this example, however, the work was related to installing economizer elements in a boiler. To maintain schedule for the overall job, the welding for this phase was started as soon as there were enough elements hanging in place to give the welders some work. Then, as more elements were hung in position, more welders were brought in, and as the elements were welded out, the number of welders was reduced accordingly.

It is important to note that although the first chart in figure 11–2 reflects the number of welds of a particular job, it could just as easily reflect any of the other categories established for the project, such as the number of electrical terminations, feet of piping, or square feet of insulation. The purpose is to measure progress and productivity, at a discrete level, so that the work can be managed in a cost-effective manner.

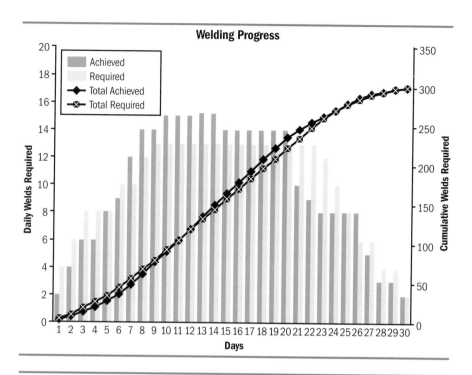

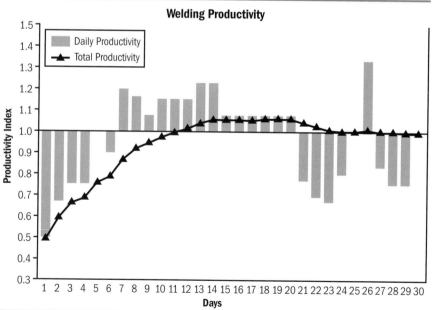

Fig. 11–2. Welding progress and productivity

Using a chart such as the one just discussed is a common method of measuring production and how it is affecting the schedule. Another method is to perform a calculation of the productivity and then graph it. When this method is used, the norm is set at 1.0 (equal to 100%). Anything less than 1.0 is a reflection of underproducing, while anything higher than 1.0 is an indicator of additional progress. The second graph of figure 11–2 shows the same data as in the first chart, except it is now in the form of productivity. As seen before, the progress or productivity at the beginning of the job is less than 1.0, or behind the plan or the norm. As the job moves forward, on the fifth day, the productivity is exactly 1.0. It stays above that until the work effort is reduced to coincide with the schedule. Note that the cumulative productivity reaches 1.0 on the 11th day and then stays above or at this number for the remainder of the job. Obviously, this is what is wanted.

Now that we see how productivity can be measured, what can be done to affect it? First, there is the skill level of the craftsmen themselves. The right people must be assigned to the right tasks. Then, just as in the earlier bridge-building exercise, there is the number of workers assigned. The right number of workers must be assigned to keep the work moving in accordance with the schedule, and that does not always mean more workers. Frequently, there are space limitations where too many bodies would actually reduce productivity. Also, there are the tools and equipment. The right tools and the correct equipment must be available to perform the work. Adequate consumables must be available. Workers standing around waiting for these items will not be productive.

Finally, there is the schedule, or work program itself. Take the case of installing superheater elements in a section of a boiler that requires refractory pours simultaneously with the element installation. The boiler-makers install two or three elements, then they come out of the unit while the refractory is poured. They go back and install three more, come out again while refractory is poured again, and so on. Not very productive. A change in the work plan and a change in the schedule might be in order to get an increase in productivity here.

There is one other measurement of productivity that is particularly relevant to the power plant construction business: the measurement of welding productivity. Because welding is often a major part of the work in a power plant, and because this work is often on the critical path of the job, measuring the welders' productivity is a way to gauge their effectiveness. However, since welders often work on different size welds, at different times, it is useful to establish a common denominator against which the welders' work can be measured.

To do this, a common practice is to calculate how many equivalent welds are contained in each weld; then measure the number of equivalent welds a welder, a pair of welders, or the whole team of welders makes in any one day or week; and compare this number with the pre-established norm. One way to develop equivalent welds is as follows:

- Calculate the amount of weld material required to weld two 2.0-inch diameter tube ends with a wall thickness of 0.25 inches each. Call this one equivalent weld.

- Calculate the amount of weld material required to weld each weld on the job, and divide this by the amount of weld material required to weld the above 2.0-inch diameter tube. This will determine the total equivalent welds.

However, since that is not such a straightforward calculation, the following method is commonly used:

- Subtract the square of the inside diameter from the square of the outside diameter and multiply by 0.5714: $(OD^2 - ID^2) \times 0.5714$ = *number of equivalent welds.*

The productivity measurement is then done the same way as before, except now, instead of counting actual welds, the equivalent number of welds is used. For example, if the tubes to be welded have a diameter of 2.75 inches and a wall thickness of 0.375 inch, these tubes would be considered to have 2.04 equivalent welds each. If the same tube has a wall thickness of 0.625 inch, its equivalent weld count would be 3.04.

Overtime

As discussed in the training exercise example above, productivity can certainly be improved with some out-of-the-box thinking. However, productivity, in and of itself, is only one part of the formula that results in work accomplished. The number of hours that the workers actually spend working is the other factor. And here the question arises: What is the appropriate number of hours a worker can be expected to work and still be productive?

Based on various studies by the author over a number of years, some factors have been developed that approximate efficiency losses when workers work beyond the standard eight-hour day, five days a week. Using the eight-hour, five-day norm as a base of 1.0, or an efficiency of 100%, these studies have resulted in the factors shown in figure 11–3. Note that the assumption is that the first eight hours are 100% efficient. It is only the follow-on overtime hours that are inefficient, per the tabulation shown.

Impact on Productivity Due to OT Inefficiency			
	1 Hour OT	2 Hour OT	3 Hour OT	4 Hour OT
5 Day Week	8%	15%	20%	30%
6 Day Week	20%	30%	35%	45%
7 Day Week	25%	35%	45%	60%

Fig. 11–3. Impact of overtime on productivity (Courtesy of Construction Business Associates LLC)

There are also several other factors that can impact work accomplished. These are absenteeism, fatigue, and overstaffing. All will also impact the actual work accomplished in various ways and should be considered when reviewing progress.

Scheduling

No power plant construction job is run without some type of schedule, even if it is very rudimentary, and no power plant owners would allow their money to be spent without a plan, or schedule, that provided some semblance of order for the execution of the work. So after labor management, schedule management is next in order of importance to the project. However, to delve into all of the ways to schedule a project, and to explore all of the different tools and methods that are available to do this, would require more space than this book can provide. The subject of planning and scheduling, like labor productivity, generates its own volumes of books. But no book on any type of construction, including power plant construction, would be complete without some discussion of the subject.

There are many different methods available to schedule a job, ranging from a hand sketch to a simple bar chart to sophisticated computerized programs. Each has its unique advantages and disadvantages. Not every project requires the most sophisticated process; some projects can suffice with a hand-drawn bar chart. However, most power plant construction projects consist of hundreds and hundreds of items, even thousands at times, and managing this many mini-projects requires sophistication.

The most common scheduling programs are commercially available from a variety of vendors. Quite often, the decision to purchase a program will depend upon the existing software already in use for estimating, accounting, payroll, personnel management, and purchasing. To be able to integrate the schedule with payroll and purchasing information and compare all of these data with the original estimate, or use them to prepare a future estimate, often justifies the higher priced software programs available on the market.

Once a software package has been selected, the next hurdle is to get it installed and populated with the site-specific data. Some organizations have full-time professional schedulers who can sit down with the original estimators of the job and build the outline of the schedule. They get together with the site supervisory team and input all of the WBSs and any other details that will be used to manage the daily activities of the project. They also work with the site accounting, purchasing, and payroll staff to integrate the job-site costs, arranging for a system that allows these costs to be input only once; for example, they may set up a central repository to which all data will be sent and from which each respective user can download information as needed.

After the schedule has been built, it needs to be reviewed to ensure that the logic makes sense and that all potential outside impacts and constraints have been taken into consideration. For example, if the schedule is for use by the general contractor (GC), it must also address the work scopes of each of the subcontractors. This can be done on a macro scale, using higher-level work groupings than the subcontractor would use. The GC's schedule must also address impacts from the owner. These could be issues such as receipt of permits to allow the work to proceed, hold-points that the project lender may require, equipment de-energizations, or access to areas that only the owner controls. Ultimately, an independent, third-party review is always beneficial. By employing a third party, any bias from the builder of the schedule will be removed.

With the schedule complete, reviewed, and accepted by all parties, its maintenance process must be established. Many projects have well-built, excellent schedules when the job starts, but then the schedule is not maintained in an orderly fashion. Too often, the scheduler is assigned duties outside of scheduling. At other times, the scheduler is a pure scheduler, someone who is a software whiz but has no idea about the difference between a steam drum and a turbine rotor. Either of these situations prevents the scheduler from being able to thoroughly analyze the data being provided for input to the schedule. Therefore, it is very important that the scheduler be familiar with the components of the work and have the time to do this job.

As an example, here is what can go wrong when the scheduling is not properly maintained. Take the case of the erection of a new boiler. Generally, as the work progresses, the field superintendents provide the scheduler with updates on the progress of the work. The superintendent responsible for the backpass may report that the reheat elements are 50% erected. The scheduler, if not familiar with the components of a boiler, will simply enter this information into the schedule. But suppose the superintendent was wrong? The foreman may have given the superintendent erroneous information or the superintendent may have just been guessing as to the percent complete, not actually having had the time to make a physical count of the number of elements installed. Either way, if the scheduler is an experienced member of the power plant site staff, the scheduler would have walked through the job before the superintendents gave their reports so that the scheduler would be able to second-guess them. In this case, the scheduler would have been able to ask how can 50% of the elements be erected when 75% of them are still in the laydown yard? The scheduler would have been able to avoid erroneous reporting, in this case, overreporting.

The opposite also happens: The field staff sometimes reports less progress than actually achieved, sometimes referred to as "keeping something in the back pocket." Just as with overreporting, underreporting is a poor way to manage a job. In the case of underreporting, cash flow may be hindered, and as discussed in earlier chapters, reduced cash flow, the lifeline of a contractor, can lead to serious consequences.

An error frequently made when setting up the job-site management procedures is not providing a single source of control for all scheduling activities. On a job with more than one contractor, it is important to have *only one* master schedule, and *all* other schedules must be synchronized with this one program. Otherwise, chaos will set in.

What generally happens in the event of nonsynchronized schedules is that the contractors, and even their subcontractors, will each build a schedule to suit their needs with minimal regard to the needs of the other contractors. For example, the electrical contractor may schedule the installation of cable tray and cables across the rear of the boiler at a point in time before the boiler contractor has scheduled the installation of the reheat elements into the backpass. If the work proceeds in this fashion, the cable trays and cables may have to be removed and then reinstalled so the reheat elements can be erected, or the erection scheme for the reheat elements may have to be redesigned. Each option is costly and could have been easily prevented if both schedules had been synchronized, and reviewed, by a single individual in charge of the project-scheduling process. For more detail on this example, see chapter 2.

Now, having addressed the importance of building a proper schedule and the importance of managing its maintenance, we must also address the importance of keeping copies of all schedule updates and revisions. This becomes crucial in the event of disputes. As discussed in the chapter on claims avoidance, when all of the schedule updates and revisions are available, the claims process goes much more smoothly. Facts are facts, and if the scheduling process was properly managed, the impacts of delays, extra work scopes, and accelerations could be readily established. Without these earlier schedule revisions available, claim managers must spend extra time and money to reinvent the sequence of events, which ultimately leads to increased costs for all concerned.

Tools and Major Equipment

Directly managing the site tools and equipment can yield major savings. Although the cost of the tools and equipment of a 14-week outage may run in the neighborhood of only 10% of the total job-site costs, on a new construction project, they may easily exceed 20%. So if the construction costs of a new plant are $50 million, 20% of this is $10 million, a sum that easily justifies direct management of this area of the job (note that on many jobs, these costs are much higher).

Many site superintendents focus most of their attention on only three aspects of the job: labor, schedule, and safety. They seldom spend much time worrying about the tools and equipment being used; this is treated as an entitled supplement for the management of the labor and the schedule. Consequently, tools, especially hand tools, are often ordered at the last minute when the stock is gone. They are also often procured and held as "private stashes," not made available to any other work crew. In the first instance, this can create unnecessary freight costs when overnight or next-day shipment is required. In the second instance, it creates unnecessary inventory, which equals unnecessary costs (fig. 11–4).

Fig. 11–4. Excess inventory of tools (Courtesy of Construction Business Associates LLC)

A similar situation occurs with large equipment such as cranes. Although the procuring or renting of a crane is usually planned in advance and is based on the lifts that need to be made, the disposition of this equipment is another matter. Once again, the site superintendents often treat this equipment similar to the way they treat their tools, as an entitlement for managing the labor and schedule. Frequently, after the main tasks for which the equipment was ordered are completed, the equipment sits idle on the job site, either because someone has a thought about using it for an additional task in a few days, or because everyone has now forgotten about it. Either way, it usually continues to incur costs by way of rental fees; or if self-owned, it incurs direct job-site charges for its depreciation.

To alleviate these unnecessary costs to the job, there are several ways to proceed. One approach is to assign an individual to manage all of the tools and equipment. This person would be responsible for making an inventory of all tools and equipment on the site. Then, an assessment of the remaining needs would be prepared, say, on a weekly basis, and the two compared. Surplus items would then be released immediately, and a plan developed to manage the shortfall. Subsequently, a review would be made of ways to substitute less costly tools and equipment for those still being used. As the job neared the end, daily reviews would be made. Even if this process only saved 1% of the tool and equipment costs, on a major job this would far exceed the cost of the person managing it.

Another approach is to subcontract the supply and management of these tools and equipment. There are various companies that specialize in this. As Pat McKenna of F&M MAFCO stated during an interview several years ago:

> *Asset management* is a way contractors can control overhead expenses. Instead of dumping a lot of money into facilities and specialized tooling, they can outsource this. Our company has developed a way to meet these needs with our "ON SITE SUPPORT SYSTEM." This system affords the contractor a method to have "What they need when they need it" by having a fully stocked warehouse on-site offering 24/7 availability, with inventory controls, managed by us. Small tools, consumables, communication devices, welding systems, air compressors, step-down transformers, tube milling/ expanding tools, tuggers and other hoisting devices, rough terrain and crawler cranes, all from a single supplier built around and for the project at hand.

In addition to managing the job-site tools and equipment, it is also important to review the rigging plans at the beginning of the job to find out if there are more cost-effective ways to perform the task. For example, figure 11–5 shows a hydraulic crane being used to erect waterwall panels in a boiler. Is this cost-effective? There is only one way to be sure, and that is to assess the expected costs for the crane rental and crane labor, and then compare them with the costs of doing the job using tuggers. It will be a function of savings in labor costs (due to the shorter time span by using the crane) plus the higher costs of renting the crane versus using tuggers and their associated costs.

Fig. 11-5. Is this more cost-effective than using a tugger? (Courtesy of Construction Business Associates LLC)

Consumables

Just as it is important to manage tools and equipment, it is equally important to manage consumables. Although not as costly as tools and equipment, consumables still impact total site costs, ranging from 5% to 10%. Therefore, reining in unnecessary waste in this area has an impact on the bottom line. When no specific plan is made to manage these items, they end up being treated the same way as the tools and equipment when they are not independently managed—they are treated as an entitlement.

There are several ways to manage consumables. The simplest is to include them in the duties of the person assigned to manage the tools and equipment. This person would then do precisely what he or she does with the tools: make an inventory of all consumables on the site. Then, make an assessment of the remaining needs, say, on a weekly basis, and compare the two. Surplus items would then be returned for credit, and a plan would be developed to manage any shortfall. This is usually adequate for a short duration outage; but what about the long-term, 18-month construction job?

Because it is difficult to plan for the use of consumables stretching for an 18-month period, a risk shift to the supplier of these items is often effective. Similar to the example statement in the section on tools and major equipment, the supply and management of consumables can also be outsourced.

The outsource contract should state that the supplier must establish a facility on-site (a trailer or maybe a fixed building) for housing these supplies. The workers could then go to this facility and draw out the consumables they need for their particular task. The supply clerk, an employee of the supplier, would be responsible for recording the cost of these items with the proper accounting code, tracking the inventory, and reordering as the stock dwindled.

At the completion of the job, any consumables remaining in the supply shack or trailer would remain the property of the supplier. Any unused supplies still remaining on the site would be returned to the supplier, for a credit, and the supplier would then remove its facility from the site. This type of arrangement has the advantage of shifting the risk of overstocking from the contractor performing the work to the supplier providing the consumables. In the event of shortages, the supplier would be responsible to supplement its on-site supplies from its other locations or stores.

Materials

Not every job requires the site team to procure and/or manage the materials to be installed. Often, this will be the responsibility of the owner, original equipment manufacturer (OEM), or GC. But there are those jobs that are delivered and erected (D&E) where the on-site contractor has the responsibility to procure, deliver, store, and erect the materials. This has its good and bad points. The good points are that the site has more control over what is being provided, when it is delivered, and often how it is stored. This makes the job of scheduling the work, personnel, tools, and equipment easier. It allows the construction contractor to work directly with the manufacturer or supplier to arrange for construction-friendly delivery sequencing. The downside of this approach is that the site management is now solely responsible for on-time deliveries, correcting manufacturing errors, and any difficulties in installing the materials. They now have to provide their own purchasers and expeditors. They may need additional engineering staff to work with the manufacturers. They no longer have anyone else to point a finger at when the materials do not arrive when promised and in the condition promised.

Since most construction contractors do not usually have personnel experienced in engineering, purchasing, or expediting, material management is generally not part of their scope. When forced upon the contractor, this often diverts site management's attention away from the daily site work activities, which can result in less effective site managing and increased costs.

Site Services

Site services are needed by everyone on-site, from the owner down to the GC and from the GC down to the contractors and their subcontractors. Everyone needs a place to park. Everyone needs a place for their offices and other facilities. Everyone needs power, water, maybe compressed air, and trash removal. Traffic control is important on a new construction site, and road maintenance may be required. Security is always a major concern. Storage/laydown areas have to be maintained. Nighttime lighting may be necessary. First aid and possibly ambulance services may be required. Sanitary facilities, fire protection, and communication facilities may also be required. The responsibilities for all of these have to be assigned and assumed.

Usually, the GC assumes these duties for a new construction project. If it is work in an existing plant, the owner is often in charge. But to whomever these responsibilities are assigned, they will still need to economize the expenditures. It is not easy to please all of the contractors and suppliers that are depending on these services and still remain frugal. Everyone wants their facilities to be located right next to their work. Everyone wants to park right next to their own facilities. Many contractors want unlimited power supplies to do their work, and many want laydown and preassembly lots within easy reach and well maintained.

The typical construction site is never designed to provide all of this. Most existing plants where an outage is scheduled are even more restrictive. There is never enough of everything to please everyone. So the owner or the GC resorts to a juggling act. Some of the responsibilities are shifted down the line. The owner of an existing plant may require the individual contractors to provide their own compressed air. They may require them to rent land outside of the plant for laydown and preassembly. The GC on a new construction project may insist that all parking be outside the premises (with the GC or owner providing buses to transport the workers from there to the job and back).

Most contractors do not like to share. They also do not want to be moved from location to location. They want to establish themselves on-site and put that phase of the job behind them. So it is incumbent upon the owner or the GC to plan how to supply site services and manage them well. But it is equally important for contractors to plan ahead and advise the owner or GC of their needs long before they are required.

When there is no coordination of site services, inefficiencies result. Welders may stand around waiting for additional power to be provided. Riggers may wait while access is prepared to allow a large component to be brought to them. Personnel may need to walk up many flights of stairs while the access elevator is being repaired. The examples go on and on. The parties responsible for providing these services must work closely with all of the end users. They must manage this as a stand-alone project. The costs of not doing so are often intangible, but nevertheless, they are real.

Keeping It All Together

Managing all of the activities on a power plant construction site is hectic at best, especially when it is a fast-track project or a tightly scheduled turn-around. There are so many different activities, so many different entities, and so many different needs that an organized process is required. Typical site managers, whether they are the owner's representatives, the lead people for the GC, or the leads for the contractor, have more duties than sometimes seem humanly possible to perform. Added to the daily emergencies that arise, the site managers have almost no time to verify that everything is going as it should, all of the time. A systemized check-off procedure is needed to maintain some semblance of order.

The design of this system cannot be complicated because if it is, it will not be used. It can be lengthy, but it must be logical and it must be organized. One approach is to follow a process similar to an auditing procedure. First, classify the project into its various business elements:

- Project management
- Administrative functions
- Purchasing
- Labor
- Safety
- Quality

Then make a list of each activity, business requirement, or document to be kept on-site. Review each of these items, and place a "yes" or "no" next to each. However, do not review the total list of items all at once. Space the reviews over a period of days or weeks. Subsequently, make spot checks using this list as a guide. (See Appendix E for an example.)

Many projects can benefit from a direct review of the items exactly as they are listed in Appendix E. All of the items will not necessarily apply to any single project. Some can be eliminated, while others may need to be added. But the importance of using such a procedure is that there is now a process that can be consistently followed to help maintain the focus on the business side of managing the site activities.

E-Commerce and the Internet

What is e-commerce, and how can it, and the Internet, facilitate good on-site construction management? As technology evolves and as companies streamline in the name of cost reduction, most organizations are using electronic media for and in their daily business. They use these media for communications. They use them for transmittal and storage of information, as well as to manipulate data. Media provide a portal for information exchange without time-sensitive boundaries. Collectively, these media are often referred to as B2B (business-to-business) exchange. But where is its applicability for a construction project?

Let's go back to the different segments of a construction project: the labor, schedule, tools, and the like. Each of these can be beneficially impacted with the right application of electronic technology. Labor recruitment can and is being done through the use of large databases that allow for searching for persons with the specific skills that are needed for the job at hand. This information is available from in-company files, labor brokers, and union halls for the benefit of the site management looking to staff a job. These databases can provide experience backgrounds, educational training, and technical skills for literally millions of workers who have the potential of working on the site. What used to be strictly a laborious word-of-mouth process of identifying workers and supervisors has become a much more refined process of searching and selecting the right person for the right job.

The process of scheduling was the first to become computerized. With the development of logical links tying hundreds of items together, the use of the computer became standard. It allowed not only the linking of items together but also the calculation of extra time in the schedule (floats), or the loss thereof, when an item's time slot or duration was changed. From this stage, it became second nature to link various schedules together electronically. This allowed different parties on the project to integrate their respective programs with the master program, producing a total project program that would highlight the impact of one contractor's actions on the others.

Taking this a step further, schedules are now updated by input, not only from the various contractors but also from suppliers of equipment and materials. For example, the turbine pedestal pour is scheduled to commence on a certain day, but the concrete supplier has a mix-up in the availability of trucks, creating a two-day delay. This can be electronically piped right into the master project schedule where the impact on all affected contractors can be immediately seen. This allows for work-around options to be explored and implemented, avoiding inefficiencies by keeping the labor productive, even if on another task. It allows for on-the-spot planning.

Tools and equipment management can be accomplished much more easily by using e-commerce. Tracking of the tools can be done by the use of bar codes. Each tool can be marked with specially formulated labels (or sometimes just engravings) that allow anyone with a handheld scanner to send information to a central database that will advise those responsible for managing the tools of the status of each one. Information on who has the tool, where they are working, and what cost codes are involved is instantaneous. If a tool is lost or broken, this can be recorded, inventory can be updated, a new tool can be ordered, and/or repair of the broken tool can be scheduled.

Major equipment management is also much easier when all of the information resides in one database. If the contractor owns the equipment, it can be tracked as to location and requisite cost codes. If a third-party rental company owns it, they can access their records and know exactly where it is and when it is scheduled to return. For either party, the equipment's maintenance can be tracked and automatically scheduled. And for the user of the equipment, its usage performance can be calculated, allowing for fact-based decisions on how to get the most for its cost.

Managing consumables is another area perfect for e-commerce, given that many of the consumable items on a power plant construction project are repetitive items, lending themselves to be purchased in bulk. E-commerce makes this process painless. With a tracking system that feeds into a central database, those responsible for purchasing can be electronically notified when the stock of a particular item is low. They can then order more (electronically if they are linked to the supplier), and the supplier can invoice in the same manner. After receipt of the items, an electronic or scanned inventory can be made, and if this tallies with the data used for purchasing, payment can be authorized immediately.

Even managing the day-to-day needs of the site services can be expedited using e-commerce and the Internet. Trash-removal services can be scheduled and updated with the service provider on a live basis. Construction power usage can be tracked and compared against projected demand to help decide when increases or decreases will need to be made. The tracking of open areas, the planning of roads, both temporary and permanent, can be made and transmitted to all parties on the site, without the need to wait for the next scheduled meeting.

But the most powerful effect that the electronic media and the Internet can have is in the business management side of the site work. The investors and the owners of power plant facilities expect timely, fiduciary management of construction work on their facilities. They want information so that they can make decisions that may affect their total investment portfolios, going beyond just the plant in question. They do not want surprises. As stated by John Long, former president for generation, Constellation Energy, in the Introduction to the original edition of this book:

> The successful power generators are those who can consistently meet and beat their forecasted earnings. Excellence in project management and project controls are vital competencies to achieve those goals.[1]

Today, excellence means being able to collect and digest data and transmit the results to the decision makers. The electronic media and the Internet are excellent tools for this.

Regarding labor management, schedules, tools, and major equipment, e-commerce and the Internet are the paths of project management that offer flexibility and control. Rolling in the consumables and materials and even including the management of the required site services just enhance this. Information technology (IT) is a vehicle for power plant owners to show their investors that their business is under control. The commonly used buzzwords are *"centralized, enterprise,* and *program control.* The bottom line is that data are collected in a central repository in a timely, instantaneous manner and applied according to the needs of the user, in time to make proactive decisions.

Interfacing is the key. There are many organizations providing many different services to a host of end users, and not all use the same platform for data collection or information dissemination. One contractor may be accustomed to using a particular software for scheduling. This contractor may have homegrown payroll systems and use a third-party program to marry the two. Another contractor may use completely different systems, grown out of the need to satisfy clients in a different industry. A third contractor may have a different variant on these processes, and then there is the GC, who is trying to coordinate them all. It can be a juggling act, at best, unless there is a program or process to connect all of this information.

Fortunately, there are programs, vendors, and often in-house providers that can assimilate all of these data and provide feedback that will meet the requirements of every entity on the job. The information from the first-tier contractors can be fed into a central repository, collated with the requirements of the GC's project needs, and superimposed on the owner's requirements to allow the parties in charge of the project to make decisions in time to impact the work. Basically, these systems work as described in the next paragraphs.

All site information, such as budget, WBSs, cost, labor hours, schedule, material control, nonconformances, request for information, and safety, are electronically stored in one central location. The path that the information uses to arrive at the central repository is not relevant; the information can come via any software program, even competing programs. For example, one contractor may be using a scheduling program that is in vogue today, while another may be using one that was in vogue the year before—it does not matter. The same is the case for information from the accounting group. One may enter information using a spreadsheet, while another may be using a database format.

A retrieval system is designed to provide information (as opposed to data) for use by personnel according to their needs. For example, the site manager might want to know the overall status of the electrical work on a new large power plant project. There may be four or five different contractors involved, one for the basic grounding grid, one for the high-voltage switchgear, one for the control room, one for the exterior cable trays and cabling to all of the equipment, and several others who are responsible for installing instrumentation and controls. Because all these contractors feed their data into a central repository, the site manager can call for and look at collated information showing current and planned progress, costs versus budget, and documented issues that either have been resolved or still need to be resolved. The system can also be designed to integrate with home office or corporate systems, so if someone in the home office wants to see the committed costs to date, or the total labor hours used and projected, they can simply access it by the click of a computer mouse.

For busy site managers, the system can be designed to provide a status report of all major projects or contractors on-site, showing budgets, expended costs and committed costs to date, projected final costs, schedule information, requests for information resolved versus still outstanding and overdue, nonconformances, and safety statistics. Another feature that can be built in is the ability to add notes to clarify or explain some of the indicators. The two essential benefits of using a system like this are overall dashboard viewing and specific drill-downs.

Dashboard. The dashboard, as figure 11–6 shows, provides summary information rolled up to any level desired by the viewer, with green, yellow, and red traffic-light-type symbols that advise the recipient of the condition of each portion of the project's status. Looking at figure 11–6, one can see that the grading project is in trouble. Both cost and schedule are highlighted in red, and a quick glance at the approved and projected funds shows why. The note at the end of the line even talks about bankruptcy possibilities, suggesting this particular part of the job requires immediate attention.

The foundations project is also worthy of attention, but for different reasons. Here, the costs are creeping up. The red warning light under the nonconformance report (NCR) column is worrisome. This indicates that there is an issue with the NCRs to the specifications that could translate into money or schedule problems as the project proceeds.

| ABC Power Project Anywhere, USA | | | Executive Summary | | | | | | Major Site Activities Summary August 2013 | |

Project	Contractor	Approved Funds	Projected at Completion	Cost	% Compl	Schedule	RFI	NCR	Safety	Notes
Grading	Morse Earthworks	$275,000	$400,000	(R)	87%	(R)	(Y)	(G)	(G)	Sub may resort to bankruptcy protection
Foundations	Municipal Foundations	$500,000	$575,000	(Y)	55%	(G)	(G)	(R)	(G)	NCRs starting to affect costs
Electrical	Ramsey Electrical	$275,000	$275,000	(G)	5%	(G)	(G)	(G)	(G)	Contractor just mobilized
Electrical	Specialty Instruments	$480,000	$480,000							Contractor not yet on site
Turbine	Self Performing	$8,500,000	$8,500,000							Activity not yet started
Boiler	Boiler Specialists	$40,000,000	$40,000,000	(G)		(Y)	(G)	(G)	(G)	Material delays may affect schedule
Site Total		$50,030,000	$50,230,000		0.5%					

Key: (G) Green = No Problems Envisioned
(Y) Yellow = Potential Problems Ahead
(R) Red = Requires Action

Fig. 11–6. Dashboard view of major site activities (Courtesy of Construction Business Associates LLC)

At the point in time of the subject report, there do not seem to be any electrical construction issues, but the boiler contract has a cautionary yellow indicator in the schedule column, and the note points out that material delays may soon affect schedule. This is the kind of instant warning that can assist site managers in determining where to focus their time and whether or not they need to drill down further to see what else may be lurking in the shadows.

Drill-downs. The other benefit of using this type of system is the ability to drill down and get to any detail, such as the labor hours expended to date for a particular concrete pour or the number of lost time accidents one of the electrical contractors has experienced, including the details of each accident. It is actually this ability to drill down that makes the dashboard so valuable. The knowledge that the data underlying the dashboard view are available, at the click of a mouse button, imparts the necessary confidence to the user that what's being seen is what's actually happening.

Some organizations go beyond just the site use of this type of project management tool; they integrate it with their corporate systems. For example, the details of a specific job, such as the one in figure 11–6, can be added to the details of many other projects, whether similar or not, to keep track of and manage total corporate resources like capital expenditures, supervision, company-wide safety statistics, overall cash flow, tools and equipment tracking, and so on. Properly designed, this type of system can be accessed from anywhere in the world; the possible uses are many. But the single key is that it is no longer just a repository for data, it is also a window into the information for those that need it the most—for them to make proactive decisions—keeping with the concept that quiet proactivity is always better than heroic reactivity. It helps projects complete sooner, at less cost and with greater stakeholder satisfaction.

To expand a bit more on the uses of e-commerce and the Internet that can facilitate the management of a construction site, just look at the handling of drawings. When they are transmitted over the Internet, an enormous amount of shipping and handling of paper documents is avoided. Drawings can be transmitted electronically to the site and printed for only those workers and supervisors who need for them. Additionally, when errors are discovered or work-arounds are required, communications between the field and the engineers can be expedited. When accompanied by live video feeds, the site superintendent and the home office engineer can correspond just as if they were standing together on the site with the drawings in their hands.

Another tool is bar coding. To speed up data gathering, many segments of the construction process lend themselves to the use of bar codes, that series of vertical lines of varying thickness that are on almost every item purchased in any store today, from groceries to clothes to automobile tires. The information that these bar codes contain can be enormous. For example, the bar code on a tool can be scanned periodically, and the site management can use the tool's cost code to charge the tool's cost to the proper WBS; update the total usage time of the tool; and then automatically schedule it for maintenance, replacement, or recalibration.

Bar coding of incoming project materials can assist in identification and material receipt inspections. Information in these bar codes can be used to determine where to store the materials, especially with regard to when the material needs to be accessed and used. For example, if the original construction plan calls for preassembling the boiler waterwalls, along with buckstays and sootblower and observation port seal boxes, it is important to arrange for the storage of each of these separate components as they arrive on-site. Using bar codes, these items can be flagged to be stored near each other, and they can also be coded so that once they are all on-site, the scheduling department is notified that the preassembly work can start.

Other uses of bar coding can extend to personnel. Bar coding their identification badges can eliminate the brassing in process still used at some sites. It can even aid in reducing the need for the foremen and timekeepers to manually input a worker's time, thereby reducing the possibility of charging the time to the incorrect cost code or WBS.

Some site superintendents also expedite the data gathering and communication steps of the e-commerce process by using handheld personal digital assistants (PDAs) and/or smart phones. Some of these devices can be used to read bar codes on equipment as it is received, and some can be used as keyboards to enter information that the superintendent wants to transmit to the central repository of data. Others can also be used as cellular phones or as walkie-talkies. The phrase "untethered communications" is used to describe this type of information transmission because these devices either store the data until they are plugged into a receptor for downloading or they transmit the data wirelessly to wherever it is to be stored.

Rugged, handheld tablets similar to PDAs are also available. These are designed to hold large amounts of data, such as drawing files, and they can then be used to update, or redline, drawings on the spot. The tablet can be programmed to constantly update a central database and other tablets as well, so that all is not lost if one tablet suddenly falls 200 feet off the building structure.

And then there is the whole new world of *apps*, or software applications. However, an app is not just any old software program—it is a special type of software program. An app typically refers to software used on a smartphone or mobile device such as an Android, iPhone, BlackBerry, or iPad, and is called a mobile app or an iPhone app. Web or online apps—software that can be accessed and used while online rather than software residing on a computer—are also used in business settings.

There are thousands of apps now available for the construction industry. They are available for almost everything. Some can instantly calculate the amount of rebar and concrete needed for a particular pour. Some allow for instant access to on-site cameras to view progress or problems. Others allow building information modeling (BIM) viewing, and some even assist in crane operator signaling.

None of the above suggests that e-commerce and the Internet are going to replace human thought. Systems break down, people make mistakes when entering data, and others err in its interpretation. There will always be the need for a live human being to look over the shoulder of the electronic transactions and provide a reality check. For more on the state of the construction IT world, and how to stay on top of it, see chapter 13.

Summary

Managing the site activities of any construction job can be an enormous undertaking. Making a change of any one of the variables impacts the others. Personnel loading affects the schedule. Scheduling changes affect the cost of the tools and equipment. Not allowing for the right amount of consumables, at the right time, can affect both the labor efficiency and the schedule. Not taking proper care of the materials received on-site can wreak havoc with the job. The management of the site services, which every entity on the job depends upon, is obviously one of the most important responsibilities the GC or owner assumes. But keeping it all together and then daring to delve into the new technologies of e-commerce and using the Internet to assist in the management of the job can reap benefits that sometimes stretch the imagination.

Labor, the heart of the project, must be managed with regard to productivity. A labor hour used is a labor hour spent; it cannot be restored. It is probably next in importance to cash flow for the contractor on a fixed price project, and it is usually first in importance to an owner on a cost-reimbursable contract. Therefore, knowing or not knowing how to track productivity and when, if, and how to react to problems can spell the success or doom of the work.

Overtime must also be managed. It can be used as a tool for regaining or accelerating scheduled progress. Obviously, an extra hour worked produces more gain than if it had not been worked. But, as discussed, overtime hours are not as productive as straight time hours. In addition to premiums that must be paid for working overtime hours, the cost of the eroding productivity of those hours must be considered.

Scheduling, which is integral with labor management, is just as important. Sending erroneous information creates all kinds of difficulties. Cash flow can go wrong, and costly rework may result. Keeping accurate records of the schedules is critical in the event of claims.

The management of the tools and major equipment cannot be overstated. Too often, these items are treated as entitlements, and when they represent up to 20% of a project's costs, even a small percentage of savings can more than pay for the cost of managing them. The same holds true for managing consumables, where there is often an opportunity to not only manage these but also shift the risk of the overall costs from the contractor to the supplier.

Site services are a shared resource that everyone on-site requires to be effective. The owner or the GC who assumed the responsibility to manage these resources must accept the responsibility to adequately and fairly parcel out these resources and not unduly penalize one contractor for the benefit of another.

Although all of these pieces of the site works puzzle must be managed as discrete units, they ultimately must be pulled together to form a cohesive project. Since even a "simple" power plant project is a colossal undertaking, any tools that can make the site managers' jobs somewhat easier should be considered. A basic checklist of all of the major business elements of the project can often reduce what seems like a nightmare to a manageable process. Using a checklist like the one in Appendix E can be a way to maintain one's sanity.

But there are even more tools and more technology that can be used to relieve some of the burden of the site manager who is trying to be everywhere at once. E-commerce, the Internet, and the tools of the electronic age, which are continuously evolving, have great potential for easing the site manager's daily duties. Setting up central repositories of data, designing systems that will distill these data into useful information, and then providing clear dashboard reports for decision making is the kind of leap that the 21st century is all about.

Bar codes and PDAs are some of the hardware tools available to assist in implementing these technological advances. They are here today, but tomorrow there'll be still more. Evaluating, experimenting, and being amenable to using new tools such as these will be the mark of the future site leaders. The need for the hands-on skills will not go away, but the tools to free up the time to allow the hands-on skills to be used will be a complement to every site management process.

Reference

1 Hessler, Peter. "Today It's All about the Money." *Power Plant Construction Management: A Survival Guide.* Tulsa, OK: PennWell, 2005, p. xviii.

Information for Decision Making 12

Most power-generation plants today are in business to make money. For them, generating power is just a means to an end. Their investors are there for one reason only, to make money. If the returns do not meet their expectations, they will shift their money elsewhere. It is vital that the management of the plant understands that. Even if the plant operates in a regulated environment, reducing costs by managing to the bottom line is still important. Investors' returns, whether regulated or not, are still *their* returns and whoever impacts these returns has an obligation to help maximize them. That means prudent management of all phases of the plant operations is of paramount consideration and this includes the often-neglected site construction phase. Put another way, it is all about the money!

As the introduction to this book emphasized, managing the construction activities of a power plant project today requires a focus quite different from that of the days of old. Equipment, technology, and operational skills have evolved—and continue to do so. Sophistication is more in demand than ever before. Not only are just-in-time deliveries important to facilitate smaller footprints and laydown needs, they are also important for the cost of inventory control and storage fees. Not only are pretrained craftsmen important for being able to hit the ground running, they are also important for the cost savings in on-the-job training avoidance that translates into saved man-hours and a reduction in schedule and costs. Not only is an emphasis on safety morally correct and an often-demanded requirement to work on many of the sites, but it is also a major factor of the bottom line labor costs. The same goes for quality—and the list goes on and on.

The management of the construction phase of a power project will impact the total costs of that project, whether it is building a new plant or rehabilitating the old. Although the construction phase occurs at the end of the project process, it is really the tail that wags the dog. The project process may start in operations, maintenance, or engineering. But then it moves into budgeting and, from there, to project management. Once, twice, or three times it is reviewed, then engineered; parts and equipment are procured, and finally it goes to the field. By then, however, the cost of a change, the cost of inefficiency, or the cost of cancellation can be devastating. The trick is to link all of the phases up front and then manage them collectively, toward a successful conclusion. The investor expects nothing less.

It is as simple as knowing that the economics of a managed process are controllable, and the economics of a random one are not. The time value of the money borrowed to build the plant, or even upgrade its components, is significant. The unavailability of a revenue stream before the plant comes on line has a direct impact on the return to the investor. The insecurity of entering into an unmanaged construction project, especially with all of the risks associated with construction work, has a cost component in the form of insurance, contingencies, and other protections. There are no silver bullets to success.

The previous chapters have delved into many aspects of the power plant construction process. Subjects have ranged from the planning process, which included how to structure the delivery of the project itself, and discussions of resourcing and contingencies to understanding the contractual commitments that were handed down. It is seldom possible for the site staff to influence the specifics of contract terms, especially since the bidding process, the negotiations, and even the final agreements usually take place long before the site managers are assigned. But it is still important that these managers understand what has been agreed upon, what has been committed to, and what they must expect and enforce. They must understand the impacts of their actions on the requirements they are obliged to carry out, such as the final schedule, price, and ultimate cost to the job or the contract.

They must understand the rules that are to be followed. Every contract today has a legal framework. It has terms and conditions that can be enforced in the courts, worldwide. If the owner or investor feels that things have gone awry, litigation may ensue. If the contractors or suppliers feel they have been injured, lawsuits will fly. There are jobs in which it seems that more time and money are spent on preparing for mediations, arbitrations, and litigations than on getting the job completed to the requirements of the original stakeholder agreement.

This legal environment has given rise to an era of risk management by opening new avenues of claims avoidance and insurance management. Lawyers are now an integral part of building a power plant, and they are even there for just a rehab job. This environment forces the job-site managers of almost all power plant construction projects to not only understand the rules of the game but also to have these rules in front of them for reference before saying much of anything to anyone.

But there is still the responsibility of setting up the site, recruiting the personnel, getting the work done, and walking out with one's head held high. How is this done, especially when at times it seems that all the odds are against a successful outcome? The site manager and the staff that support this function have the tools available to operate in today's world of litigation, distrust, ruthlessness, and constantly changing parameters. But they may need to learn new techniques and new philosophies to apply them. However, one thing has not changed—it is still construction management, not rocket science.

The business of managing a power plant construction project still requires satisfying the powers that be. These powers may not be the same owners that were around in the past, but they are still owners—owners interested in their money and its safety and the return that they will see. The shift that is permeating this industry is that many of these owners are not interested in megawatt-hour generation, they are interested in dividend dollar generation—manage the plant works to generate profits. In other words, manage the construction activities as if the money being spent on this work was coming from one's own pocket.

Many ideas, processes, and examples have been used throughout this book to address the various phases of managing the construction of power plant projects. They each have their own values. They each can be applied in many circumstances. But when put together, in a preplanned format, the information provided multiplies by several factors. The information that results from all of the well-thought-out data can influence decisions to completely change the course of a project. Pulling together these data and presenting them in informational reports will allow the site management, the operations management, and the investors to make informed decisions about whatever issues may arise.

What comes next is an orchestration of the information required to pull it all together, followed by a discussion of the importance of communicating this information up and down the organization. Finally, there is a brief discussion on the importance of the timeliness of this information and how it can impact other projects and endeavors.

The Decision Tree

To enable the various levels of management to make informed decisions during the progress of a project, many different sources of information must be tapped. A well-designed construction job will have a series of reports, all existing for the purpose of providing information to show where the job is, where it may be headed and where opportunities exist to change its course. Figure 12–1 showcases a decision making tree in the form of a job-site reporting hierarchy. At the top sits the decision maker. At the disposal of this decision maker are financial reports, progress reports (discussed further below), and progress information relative to the financial position of the project. With these three informational reports, the decision maker can readily see where the job stands and how it is trending in time to avoid surprises at the end. That is not to say that sometimes the course of the job cannot be changed, but at least the issues will be recognized and not come as a shock at the end.

Fig. 12–1. Job-site reporting hierarchy (Courtesy of Construction Business Associates, LLC)

Financial Status Report

First, there are the financial status reports. As discussed in chapter 10, there are a lot of financial data available. It must be collected and reported in a fashion that results in information, and it must be useful information. A site manager generally has a budget to pay for the job, and this budget is generally subdivided into categories similar to those used to estimate the total cost for the job when it was first being structured. The first column of figure 10–1 in chapter 10 contains such a budget. It tells the site manager how much money has been planned for labor; for supervision and accounting support; for small tools, consumables, equipment, and materials; and, in this specific example, even for specialty subcontractors and insurances.

The next column has the "expended to date" information, which may be the first indication of how the job's financials will look at the end of the project. But by no means is this the whole story; there are still many unknowns. For example, this column does not include money already committed but not yet expended, unless this is the final report of the project. There are usually issues with timing, where there are expenditures outstanding that have simply not been recorded. There is often the problem of vendors and contractors not issuing invoices in a timely manner. But it is an all-important first base of data gathering that will be necessary to build the overall outlook for the job.

The next column, the "projected at completion" column is the one that tells the story of at what cost the job is expected to complete. But how does one get to it? That requires several things. First, it requires an approximation of outstanding expenditures that have not been recorded. To that must be added outstanding invoices. And then, a projection must be made of the costs of the project to the end of the job. This projection requires more specific knowledge of where the job stands in relation to schedule, productivity, and man-hours expended, as discussed next.

In chapter 11, both schedule and productivity were discussed. But by also adding man-hour expenditure to this discussion, one then has all of the information that will be required to complete the "projected at completion" column of the financial status report.

The first piece of information needed to project the cost to complete the work comes from the schedule. Most schedules have a mechanism for tracking days allocated to an activity against days remaining, similar to that shown in figure 12–2a. This recreated schedule format shows the activities of one area of a boiler erection project. Each activity has an associated original duration of days (or maybe hours or shifts if it is an outage) and a remaining duration to completion. It also has the percent complete associated with each activity. In this example, the first five activities seem to be fairly well on track, all being complete. However, the sixth one, Activity A 1239, "align and weld furnace intermediate rear wall to upper wall," seems odd. It shows only 25% complete but over one-third of its original duration already used. And the chart also says very little about the next two activities, only that they have not yet started. Is this per plan, or is there a problem here? To find out, the rest of the schedule is needed, with the early start and finish columns as well as the total float. This is shown in figure 12–2b.

Activity ID	Activity Description	Orig Dur	Rem Dur	% Comp	Early Start	Actual Finish	Total Float
A 1234	Raise furnace upper rear buckstays	3	0	100			
A 1235	Raise furnace intermediate rear wall	1	0	100			
A 1236	Attach furnace upper rear buckstays	6	2	100			
A 1237	Raise furnace intermediate rear buckstays	3	0	100			
A 1238	Install furnace lower rear wall	5	0	100			
A 1239	Align and weld furnace int. rear wall to upper wall	13	8	25			
A 1240	Raise furnace lower rear wall panels	1	1	0			
A 1241	Align and weld furnace lower rear wall to inter. wall	6	6	0			

Fig. 12–2a. Typical construction schedule duration format

Activity ID	Activity Description	Orig Dur	Rem Dur	% Comp	Early Start	Actual Finish	Total Float
A 1234	Raise furnace upper rear buckstays	3	0	100	12 May	14 May	0
A 1235	Raise furnace intermediate rear wall	1	0	100	15 May	16 May	–1
A 1236	Attach furnace upper rear buckstays	6	2	100	16 May	22 May	2
A 1237	Raise furnace intermediate rear buckstays	3	0	100	26 May	6 June	0
A 1238	Install furnace lower rear wall	5	0	100	29 May		–3
A 1239	Align and weld furnace int. rear wall to upper wall	13	8	25	5 June		–2
A 1240	Raise furnace lower rear wall panels	1	1	0	24 June		
A 1241	Align and weld furnace lower rear wall to inter. wall	6	6	0	25 June		
						Total Float	–4

Note: Work scheduled for 5 days per week   Update 13 June

Fig. 12–2b. Typical construction schedule duration format, completed

Figure 12–2b shows the same work at the same point in time as is shown in figure 12a, but with additional information. Here, one can see when the activities were originally scheduled to start, as of the last reporting period, and their actual status as of this reporting period. Note the following:

- By adding the total float for the work of the first four activities, one day was gained.

- The fifth activity had a problem, three days were lost, resulting in the total work being two days behind schedule.

- The sixth activity, A 1239, seems to be rapidly losing ground. Five days have been used and only 25% of the work has been completed, resulting in two lost days of float, bringing the job total to four lost days.

To sum up, this portion of the project seems to have lost four days. It is not yet complete, and the welding of the intermediate rear wall to the upper wall seems seriously behind schedule. How will this affect the end date? To determine this now requires a look at the productivity analysis, the next tool that was discussed in chapter 11.

Figure 11–2 showed a typical welding progress curve and its related productivity information. Reproduced here as figure 12–3, and populated with data from the above example, quite a different story appears. Contrary to the information gleaned from the data provided by the schedule update, which shows that Activity A 1239 has already used 38% of its allotted time (five days out of 13) yet only 25% of the welding is complete, progress is essentially as scheduled. The daily welds required, as well as the cumulative welds, are essentially on target. Although welding work will always bear watching, most likely this activity will be completed without losing any time, suggesting that the two days of float shown in figure 12–2b as lost to this activity are most likely not lost.

Next, referring to the productivity graph in the same figure, it is clear that the total productivity index is almost at 1.0, where it needs to be, and although the daily productivity fluctuates, it is the cumulative progress, or productivity, that matters. This example clearly demonstrates that the job progress and the story being told by the various bits of data about the progress and probable outcome always require thorough analysis.

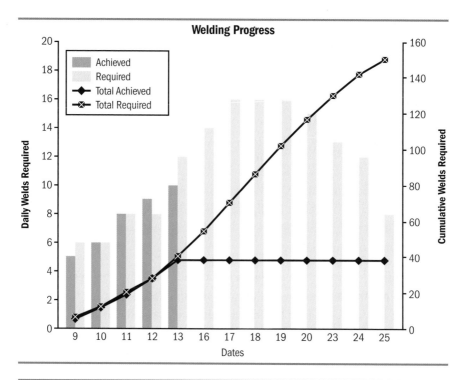

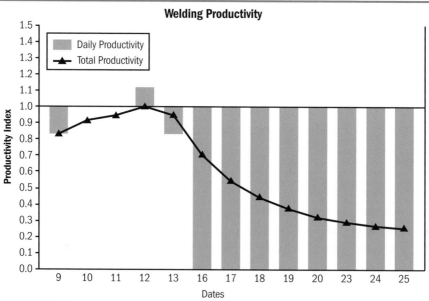

Fig. 12-3. Welding progress, and productivity

The scenario above shows that for this particular part of the project, the welding work currently underway will most likely continue to track according to plan. This suggests that the original 13 days allocated to this activity will suffice, and no additional schedule or monetary reserves will be required. It does not, however, overcome the fact that the total job is still a full two days behind schedule, as it was when the welding work started. There is nothing in the schedule to indicate that this will change, so barring additional information, the site manager must reflect a two-day negative float and its impact on the schedule and cost in the next progress report.

When a job deviates from its schedule or cost, these deviations must be reflected in the job schedule and the financial status reports as soon as they are known. Continuing with the above example, the current state of this job seems to be a potential two-day delay. If this was a short turn-around project, this delay could have a significant impact on the project's outcome. Not only might two days' worth of extended overheads, equipment, tooling, and supervision costs be significant, the loss of two days of revenue generation, especially during peak demand, could be significant. The sooner management is aware of this potential delay, the sooner alternate arrangements can be made for obtaining power from other sources to meet committed supply and mitigate the need to buy power for redistribution at high-cost spot prices.

However, if this delay is part of a much larger, long-term project, there may be opportunities to correct the deficiency. On a long-term project, personnel can be reallocated to help make up lost production. Extra shifts, or even extra hours per shift can be employed. But that does not eliminate all of the costs associated with the delay. Shifting personnel takes these workers away from other work activities that will suffer. Extra shifts or additional hours per shift will cost in terms of premium time.

To correctly reflect the impact this two-day delay will have on the job, the additional time on the short-term job must be reflected in the schedule update. The additional costs to support this time must be shown as part of the "projected at completion" section of the financial status report, possibly causing a projected overrun in the budget, which will then be a signal that more money may need to be allocated to complete the work or other scopes may need to be reduced or eliminated.

If the delay is associated with a longer-term job, and if it can be absorbed into the existing schedule, its cost impacts should still be reflected in the "projected at completion" column of the financial status report, possibly leading to a similar overrun scenario that may need special attention.

Rounding out the information necessary to complete the "projected at completion" column is the rolled-up total of the man-hours projected for each activity, along with those already expended. As part of the analysis, after knowing the duration expected from the schedule analysis, the productivity data to date can be used to assist in estimating the total man-hours required to complete the activity. This information is then entered into the financial report, converted to dollars, and added to the remaining categories to arrive at the financial projection for the project at completion.

Progress Graph

By following the above process to complete the financial status report, the information will be available to complete the progress graph. As discussed in chapter 10, this progress graph is a very handy tool to quickly see where the project is, where it is trending, and if there is still time to react in the case of problems. Plotting this graph is straightforward. The man-hour expenditure data are available from the man-hour updates used to prepare the financial status report. The actual percentage of completion data are available from the schedule update information. The budgeted man-hour data are available from the financial status report, as part of the "budgeted" column, and the planned percent complete information is available from the schedule.

This progress graph provides management with an instant view of whether or not the site work is trending in the desired direction. A quick glance can point to potential problem areas before they occur, often in time to take corrective action, and even when action is not possible, at least there will be no surprises.

Weekly Progress Report

To put a face on the information in the reports discussed so far, a written progress report is often prepared. This is usually a weekly report, ideally not more than one page long, with a series of one-line statements regarding the status of the more important activities of the job. It usually highlights one special event of the week and then has several pertinent photographs of ongoing work. There are a few simple statements regarding any problems that are affecting or will affect the work. Sometimes, it will also include a brief statement or two about the plans for the next week or two. It is not intended to replace any updated schedules or any look-ahead plans. These are still required as a normal part of the job planning activities. Figure 12–4 shows a typical report.

Weekly Progress Report

Any Utility
Anywhere, USA

Week Ending 31 March 2013
Weather: Sunny, 85°

Highlight of the Week

Steam Drum erection

Progress Photos

Duct erection in progress

Reheat Element preassembly

Progress This Past Week
1 Steam Drum raised on schedule
2 Welding FBHE Final SH Elements
3 Ground Assembly Economizer elements
4 Ground Assembly RH Elements
5 Weld membrane of Furn. Up sidewalls
6 Weld membrane of backpass rear wall
7 Attach Buckstays on backpass rearwall
8 Offload contract material arriving from port
9 Installing FBHE Final SH Elements
10 Removed Drum Lift equipment from unit
11 Preparing double wide panel transport trailers
12 Installing Duct in unit
13 Setting Stress Relieving equipment on unit
14 Setting welding machines on unit

Problems
1 No Hanger Rods for Duct erection
2 Transport Trailers arrived late
3 Buckstays misfabricated
4 Welders leaving job for nearby OT work

Safety
No OSHA recordables this week

Progress This Past Week
Planned   19.5%
Actual   18.5%

Equivalent Welds Complete
Planned   1,250
Actual   1,315

Fig. 12–4. A typical weekly progress report

Other Reports

The second tier of reports in figure 12–1—those reflecting schedule updates, productivity, and man-hours—are important, but they do not provide much information on which to act. They are there for data collection, which will be used in preparing the first tier reports. In fact, as was seen with the schedule update above, taken by itself, some of the data could have led to the assumption that the job was starting to slip further behind.

However, there are two other reports that provide both data and information by themselves. These are the quality report and the safety report shown on the third tier of figure 12–1. Both have been described in their respective chapters on quality and safety, but it bears repeating that the information from these reports is very useful for preparing the three main reports—progress, financial, and weekly.

The quality report can point to areas of rework or pending rework that may not be apparent to the schedulers when they update the schedule. If there is a lot of rework pending, for example, equipment misalignment issues, this needs to be taken into account when the schedule is being updated, and it subsequently must be reflected in the updated financial report. A similar situation arises with the information from the safety report. If it shows that safety is slipping, there could be a productivity issue looming that needs to be considered when updating planned productivity charts for the various activities.

A quick word on long narrative reports—while they are nice to have when researching what happened on a job after the fact, they are time-consuming to write and very subjective in content. Very few job-site personnel have the time to write them, let alone read them. For a very large project, they may be worthwhile, or even required by various lending institutions, but then they should be prepared by someone other than the site staff. Home office personnel, familiar with the job and its conditions, are usually better suited to address these types of reports.

Communicating Up and Down the Chain of Command

Reporting is a must on any project. Just as everyone has a boss, everyone must report to someone on what they are doing. On a power plant construction site, there are many levels in the chain of command. It starts with the worker, the craftsman who does the actual work. From there, the next link in the chain is the foreman, the one responsible for assigning the work to the workers and ensuring that the work is done. The following level is usually one of the general foremen, or *the* general foreman, who, in turn, reports to the superintendents on the job. At this point, the super-intendents are responsible to their site manager for the work that is being done, and they are responsible for keeping the site manager apprised of the conditions of the work, its progress, its costs, and any potential problems that may arise.

Beyond the site organization, whether it is the organization of a contractor or the organization of an owner self-performing the work, there are the stakeholders. These are the people who make the decisions that need to be made when the work is not going as planned. They may be higher-level managers within the power generator's organization. They may be members of an alliance that has been formed to manage and operate the plant, or they may be an independent group that represents the ultimate owners of the facility. But whoever they are, they must have accurate reports on the status of the construction activities at site.

The requirement to provide accurate reporting *up* the chain of command has been reiterated many times throughout this book. The stakeholders need good information to make determinations that will affect their money. But what about reporting *down* the chain of command? Is there a need to keep the personnel on the lower levels of the command chain informed of the plans and actions of those from above? Most modern-day management theories say yes. They say that an informed worker is an involved worker; understanding the "why" of an instruction is what will motivate the worker. Certainly, there is very little to be lost by keeping everyone informed of the decisions that are being made by the powers that be, and there is also little to be lost by providing the reasoning for those decisions.

There are organizations that share everything, everything down to the last dollar being committed on the job and the effect every change has on the bottom line of the project. Some even go so far as to divulge the cost of each of the participants so anyone wanting to calculate the potential impact of a decision will have the tools to do so. There are other groups, however, that do not go so far. They are concerned about information that will leak to their competitors and even to their clients, whether these clients are the current ones for the specific job at hand or clients from past or for future work.

The issues involved in reporting results up the chain of command are different from those involved in reporting results down the ladder. To report meaningful information, it is necessary to understand what will be done with it. Since information can be presented in an unlimited number of ways, for it to be effectively used requires that it be tailored to the needs of the recipient. The recipients at the top will use the information very differently than those below. For example, the contractor's home office manager, to whom the site manager is responsible, may need to know as early as possible if additional supervisory personnel may be required in the near future. This means that the site manager must emphasize staffing and the financial impact the lack of staffing may have on the work.

On the other hand, the client does not worry about the contractor's staffing but does worry about the effect of the contractor's activities on the other contractors on-site, an issue that the contractor's home office will not be very concerned about. Then there are the other contractors on-site. They are concerned about every other contractor's activities as well as the plans and actions of the general contractor and the owner. And what about the workers? What do they care about? Just ask them. They care about a safe workplace. They care about a financially solvent employer. They care about their reputation—being known for doing a good quality job. But for them to know the status of these issues, someone must communicate information to them.

Are the safety reports distributed to the craftsmen that actually do the work? Do they get to see how the job they are working on is rated compared to the goals of the project and compared to the standards of the industry? Do they get to see if there are financial problems with the job, their employer, or the investor/owner? Will they be able to glean whether or not the job will shut down in a week or so due to the insolvency of one of the parties—and therefore they had better start looking for another job right away? Do they get to see if the work they are doing meets the quality requirements that are expected by contract and by generally accepted industry standards? Will they have a reputation of having worked on the job that could "never get it right," or will their peers look in awe at those who completed the work ahead of schedule, under budget, and without a single nonconformance? It all requires communication.

This is not to suggest that the exact same information is provided to everyone. Of course, there are confidential matters that cannot be disseminated; some things are restricted by law. Sometimes, certain financial specifics must remain in-house. But there are many ways of portraying the results and the status of the job without divulging the details. For example, using percentages instead of actual numbers will often convey the information desired and still protect its confidentiality. Let's say that the actual costs of the work are to be held in confidence. That does not preclude management from saying that the job is 65% complete but only 58% of the budget has been spent—therefore congratulations are in order for everyone involved. The same principle applies to safety results. Often, the exact numbers are withheld due to legal or other reasons, but the percentages can be released; for example, "We have achieved an OSHA recordable rate of 95% of target!" The list goes on and on. However, even if actual numbers are not confidential, the audience should still be considered. Sometimes the information sent to upper management does not convey the message that should be sent to the workers, and vice versa. The reports that go to upper management about the details of the delays with the transportation of the turbine rotor back to the site may not mean much to the site workers, but knowing that the rotor will not be arriving for another two weeks does; to the mechanics this may mean that they really can take their kids fishing for a few days since there will be some downtime.

The important thing to remember is the audience, those for whom the report is intended and what they will do with it. The best course of action is to ask each level in the chain of command what they want to know and why. In this manner, the reports can be tailored to suit the purpose of the recipients, and a clearer channel of communication will have been established.

Program (Enterprise) Control

In today's world of doing more with less, such as making decisions with the scarcest of information, it is important that the interrelations among projects are considered, such as those among resources like money, personnel, tools, and the like. The job site manager must be aware that the decisions made on a particular project may affect other projects. Therefore, it is often helpful for the managers of various job sites within one organization to be aware of the status of the other sites; some reports should be made available to the managers of all job sites.

Personnel availability is often an issue. When several major projects are in progress at the same time, skilled labor can become a scarce commodity. Frequently, it is the site that offers the most overtime pay that gets the best workers. To minimize disruptions, coordination among sites is important. A power generator with multiple sites in one area should always be aware of the demand on the local labor pool. When several outages are scheduled at the same time, which often happens due to the short periods of time when units can be off-line, the smart site managers coordinate labor needs. They keep each other informed about the hours they plan to work, in order to avoid personnel hopping from job to job, chasing the overtime premium. They will also coordinate the labor skills they will need. For example, when one site is scheduling major welding work, the other site will not. The other site manager will attempt to schedule the major welding work either before or after the previous site requires welders.

Similar coordination is important when a contractor is planning major equipment usage, especially if it is owned by the contractor. If there are several new construction projects planned in the same area, the availability of heavy lifting equipment may become critical path. The contractor's site managers should keep each other apprised of their intentions to use this equipment so they all do not need it at the same time. Just by sharing job-site schedules, each can see what the other is planning and use this information to coordinate the use of this equipment. For example, if a contractor on one site is planning to set the generator on its pedestal in the third week of a particular month, and a contractor at another site needs some of the same equipment to raise a boiler drum, they would do well to stay in touch about their needs to avoid having to rent duplicate equipment.

One of the most important resources to be coordinated, whether by an owner or contractor, is cash. An owner must make cash available to pay internal and external suppliers across the total organization. To do this, the owner needs a consolidated cash flowchart that must be kept updated. On the one hand, the input to this chart comes from the project's suppliers at all locations, based on their anticipated plans. On the other hand, the input comes from cash sources that could be revenue generation, internal cash draw down, or lending institutions. There has to balance to ensure that (1) there will be enough cash available to pay the suppliers and (2) that there will not be an unexpected excess of cash.

If it is a contractor, especially one active on multiple sites, cash flow is equally important. The contractor must pay his labor and his suppliers as well, at all of his sites, and he must be aware of the cash requirements required to do this. But he also needs to balance these requirements with the expected inflow of cash, to avoid the need to borrow money to cover shortfalls or to avoid unnecessary cash buildup.

Coordinating these resources can be done in many ways. The simplest for the site manager on any one site is just to talk with the other site managers. However, when projects are large and complex, this is generally not effective. Often, there are so many other urgent issues that communicating beyond the immediate site is not a high priority in the mind of site managers. They do not always have the time to focus on issues related to off-site activities, even though these activities may eventually create problems for them—problems like no money in the account to fund payroll, specialized equipment not available requiring expensive third-party rentals, or local labor not available requiring travelers and the resultant costs for their travel and living expenses.

To assist with the dissemination of information that may be useful to all parties in an organization, various tools can be used. One such tool was discussed in depth in the "E-Commerce and the Internet" section of chapter 11. That tool was an executive summary dashboard. All data generated during the course of all projects can be entered into a central repository from which information can be assimilated based on the needs of the user. Referring to the dashboard shown in figure 11–6, columns could be added for personnel availability versus personnel in use, by individual crafts or by total personnel. Additional columns could be added for specialized equipment availability versus equipment that is in use or scheduled to be used. Most importantly, a column reflecting cash in versus cash out could be added.

Figure 12–5 is an example of such a view for tracking labor across an enterprise or area. It shows three different labor skills, where they are available, and the total available for the work in the area. Although in this example the supply of mechanics and electricians seems to be adequate, there is an issue with the availability of welders. Local 1 will only have three welders available at the end of the month, and since that is below a predetermined threshold, in this case 5%, the red button is illuminated, signaling that action is required. Local 60 seems to be only in the cautionary mode, but the total number of welders available from both union locals is still below the 5% threshold. Anyone with a need for welders, who also has access to this report, will instantly see that there will be problems. They can drill down by clicking on the appropriate button and find out which projects are planning to use these welders and for how long. With this information at hand, plans can be made to either rearrange schedules or start looking for welders outside of the immediate area.

ABC Power Project Anywhere, USA		Executive Summary			Labor Availability August 2013	
Skilled Crafts	Union Local	Total Labor Pool	Projected Use at Month End	Percent Available	Availability	
Welders	Local 1	94	91	3%	(R)	
Welders	Local 60	75	67	10%	(Y)	
Total		169	158	6%	(R)	
Mechanics	Local 118	111	62	44%	(G)	
Electricians	Local 134	266	31	88%	(G)	

Key: (G) Green = No Problems Envisioned
(Y) Yellow = Below 10% Availability. Potential Problems Ahead
(R) Red = Below 5% Availability. Requires Action

Fig. 12–5. Example of an executive labor availability summary dashboard (Courtesy of Construction Business Associates LLC)

Since most organization management would be uncomfortable allowing just anyone unlimited access to these reports, restrictions can be built into the program that limit access to specific individuals and limit the information they are allowed to access within the reports. Built with some foresight, "instant alert" reports could be automatically delivered to individuals who need to be aware of pending problems. In the case of the above example, a copy of this specific report could be automatically sent to each site manager who has projects in the area, as well as the home office labor coordination group. In this way, those who need to be aware of this issue would find out about it right away. Otherwise, they may not become aware of the problem until they happened to review the report during their regular review cycle.

Summary

It bears repeating that most power-generation plants today are in business to make money. Generating power is often just a means to an end. The owners and investors are there for one reason only, to make money. If the returns do not meet their expectations, they will shift their money elsewhere. For them to feel comfortable that their investment will continue to meet their expectations, they must have information about the activities going on at the plant, especially the activities during major construction work. Obtaining this information requires that the site management provide timely and accurate reports that can be used by the decision makers.

The three basic pieces of information that are usually required are progress, schedule, and cost. Using a blend of these three, a good picture can be painted of the status of the work. However, there is more that needs to be known—the future. Although there is no such thing as a 100% guarantee of accurate predicting, there are tools that can assist in determining the outcome of the job. Used early enough in the project cycle, these tools can point to potential problems, they can suggest what to do to mitigate them, and they can suggest when to do it. These tools are the graphs that depict trending and productivity.

The first tool, the progress graph, compares project completion with the cost of the work. It allows the user to see if the job is proceeding as planned and if it is within budget, and it also allows the user to determine how the results of the moment are trending: Are they trending toward a successful completion, or are they trending toward trouble? The data used to generate this information are supplied by the schedule updates, the productivity graphs, and the man-hour updates. It is important to use all of these data together because any one piece, by itself, has the potential for leading to erroneous conclusions.

The second tool, the financial status report, compares the total cost of the work completed with the budget for the job. It also contains a projection of where the costs may be when the work is complete. The data used to prepare this information also come from the schedule updates, the productivity graphs, and the man-hour reports. However, the projections rely on information developed with the progress graph.

The third tool, the weekly progress report, provides an overall picture of what is happening on the job. It points out actual progress and it advises about issues that are either current or pending. Prepared with some photographs, it provides an understanding of how the job is progressing.

Finally, the quality report and the safety report are tools that often unmask issues that are below the surface but may impact the project. When quality starts to suffer, rework will soon become an issue that directly impacts the schedule and the costs. When safety starts to slip, productivity will suffer, which again will have an impact on both schedule and cost. Although these two reports are listed at the bottom of the hierarchy of reports, their importance should not be diminished.

Concurrent with preparing reports, thought must be given to their recipients. Management above the site has needs that are very different from the workers and management on the site. Top management, and owners and investors, need information that speaks to additional support the site may require. They need to know if more supervision, tools, equipment, and money are required. They also want information to help them coordinate with others that are involved, such as other contractors on-site.

The site staff and workers, on the other hand, have different needs. They are not in a position to bring support from the outside, but they are in position to affect changes on the inside. They want to know that everyone is working safely. They want to know that they are providing quality work. They want to know if they are on schedule and if disruptions in the work are expected. They want information that helps them do the best job they can.

There is also information that needs to be shared with other parts of the enterprise, with other job sites, and the home office to assist in the coordination of the overall corporate efforts. This is information such as expected drain on personnel, expected need for specialized equipment, and cash flow requirements. If the owner or any of contractors have more than one project underway, they need to share information about what resources are available where, and what resources will be required when. Working together to reschedule events may avoid duplication of efforts and unnecessary rentals of equipment. It is also important for leveling cash flow.

Sharing information across the enterprise requires preplanning. It requires that a process be put in place to deposit the data from all sources and then to generate *information* from these data that can be used to streamline the overall operations. One convenient method of providing this information is to present it in a dashboard fashion. This will avoid information overload, the classic reason why many people never use the information they have at hand—because it overwhelms them.

In summary, there is a great deal of data generated on a power plant construction site. Some data come from the owner. Some data come from the contractors. Other data come from third parties, but all the data reflect some of what is happening at the site. The challenge is to put these data together in a way to generate information that can be used by the various parties for decision making. Not all of the information is needed by all of the parties all of the time. To avoid inundating people with information that has no bearing on their part of the project, the reporting process should be designed to provide only what is needed and to whom it is needed. When pulled together in a well-thought-out manner, the decision makers responsible for the project will have the information at hand in time to take proactive steps that may still change the trend of the job before it is too late.

Technology and the Field 13
(with content from Mark Bridgers)

This, the last chapter of this book, is an untried effort. At our workshops, my associates and I talk a bit about e-commerce and a few of the technological tools in use today, but we do not dwell on it, mostly because it comes up at the end of the day. However, the subject of using technology in today's power plant and other industrial construction management worlds is the way of the future. Therefore, we (the author and other contributors) would like to make this chapter a real go-to chapter that not only addresses what is available today (and was not available yesterday) and how it enhances managing the site work, but also how to stay on top of new and upcoming technological tool developments in the future. Our objective is to offer the reader a structured way to think about the use, application, and procurement of various technologies to accelerate the process of managing construction work. Much of what follows was not even widely available when the first edition of this book was written. But times have changed. See figure 13–1 for an illustration of the long-term change.

The construction industry is historically slow to adapt to technology relative to other industries. Therefore software and other technological systems have always been slower to move out to the construction field than to the manufacturing plant. Even though technology solutions were brought into the corporate office early on and used by accounting and operations departments to support the capitalizing and expensing of costs associated with heavy industrial construction, plant upgrade, outage, and operations and maintenance (O&M) activity, the construction side of the business never embraced these supporting technologies As these applications were developed and deployed in the office and manufacturing environments, the automation gap existing between the field, the plant, and the corporate office became more apparent.

Fig. 13–1. From the abacus to the tablet

But finally, technology *has* now started invading the job site, and mostly in a good way. From webcams to tablets (see fig. 13–2) and smart phones, various technology platforms have transformed construction sites. While field staff is still instrumental in building and maintaining various infrastructures, the adoption of newer technology allows them to do so more efficiently and effectively, and with less risk.

This chapter deals primarily with technology solutions that impact the field. Some are used directly in the field, and others are primarily used in the office, but clearly they have an impact on resources in the field, whether people, tools, materials, or equipment. This chapter will cover some important concepts to help the reader understand where technology is today and where it will be tomorrow. Finally, we consider how to select a system, how to implement it for success, and how to stay on top of the next newest system, tool, and process.

Fig. 13–2. Using a tablet to enter job-site data

Value of Systems

The first relevant question when looking into using modern technology to help manage construction is that of value. In other words, where does the value of systems and technology come from, relative to the construction of power generation assets through plant construction, plant upgrade, outage, and/or O&M activity? Some examples are risk mitigation, efficiency gains, and information flow.

This is not simply an academic consideration. Companies cannot afford to implement all types of systems without regard to the return they provide. No company should be purchasing and implementing a solution without first considering where the value comes from. It is common for a system or solution to require training and process change in order to provide its purported value. It is for this reason that a company should focus on where the value of the new system is going to come from and understand it well. From there, a good decision can be made and appropriate resources applied to the implementation.

Uses for Information Technology

There are a number of different technology classes available in today's informational technology systems marketplace. Scheduling, project management, change orders, cost control, estimating, and mobile solutions are among them. When starting out, it is best to consider how the systems and data should work to provide the risk mitigation, efficiency gains, and information flow needed. It is very easy to buy software, but much harder to get good value. Rather than simply buying solutions as needs arise, it is important to determine the impact these solutions will have on the overall construction program. Even within a class of software, there are varying degrees of sophistication and cost, from the fairly inexpensive and easy to the more advanced and expensive. It is important to know the needs and the technical savvy of the project personnel expected to use these tools as well as the support available from the selected service providers. Something too advanced could easily prove beyond the capabilities or needs of the people or organization using it.

Many software solutions available for the power plant construction industry have overlapping functionality, and as a result, the categories they belong in have become less clear. This ambiguity started with project management software, which typically handled functions like submittals, transmittals, meeting minutes, and punch list items. Early on, this was a stand-alone function and did not integrate with other software products. As the use of the Internet exploded, collaborative project management software solutions emerged and were transformed. This was the first time multiple parties—owners, engineers, accounting, and finance personnel, along with support staff—could work within the same software solution for a single project.

But just as with enterprise resource planning (ERP) or accounting solutions, the project management software solutions became large and more generic, and often times they did not specifically address needed functions. To fill this gap, "point solutions," so named because the software focused on a single function with great success, began to appear in the marketplace. With the boom in these different types of software solutions, the construction software industry began to offer different solutions to do the same function. Many times those products crossed boundaries and were classified differently. For example, there were ERP products in which the submittal (including shop drawings, material data, samples, and product data) capabilities and project management software solutions were generic as well as applicable to specific needs.

Another classification of software that evolved is the field-data capture solution. This software runs on a specific device or, in some cases, is device independent. The applications may be designed for internal crews, contractors, supervisors, superintendents, foremen, and inspectors in the field to collect commonly managed data such as time cards, equipment time, and production units. These are usually collected, routed for approval, and then sent to another system, such as ERP, for import and processing. The vendors who develop these products typically handle the integration back into the accounting software.

Drawings have been an integral part of construction job sites since the first engineer put pencil to paper. Although the process for developing and managing these drawings has changed considerably, they remain a vital and often dynamic part of any construction project. Access to the most current drawing sheet and related specification is crucial, particularly as the speed and frequency with which revisions are being made is constantly increasing. The need for collaboration on drawings is also growing, and the ability to view the drawings while maintaining the integrity of the drawings is of utmost importance.

Building information modeling (BIM) is making its way into the field as the technology moves from traditional two-dimensional (2D) paper drawings, past three-dimensional (3D) auto computer-aided design (CAD), into four-dimensional (4D) where individual 3D CAD components are linked to time or schedule-related information, all the way to five-dimensional (5D) where cost information is integrated with 4D components. Tools that work with the digitized version of a facility allow for personnel in the field to see the sequencing, context, and elements of the facility before they are actually put in place, thereby helping internal crews, contractors, supervisors, superintendents, foremen, and inspectors to plan and execute better.

Fueling, maintenance, and management of heavy equipment at the job site have been impacted by technology. Many firms have put Global Positioning System (GPS) devices on their heavy equipment, initially in an effort to cut down on theft. In this context, once the jobs site's perimeter is known, a GPS device can be set for alert in the event the associated equipment moves outside of the perimeter. But heavy equipment can also have telematic devices implemented. These work within the equipment and automatically send important equipment-operating data back to another system, either an ERP system with an equipment application or a stand-alone application, for example, for maintenance management. This can help avoid costly repairs by alerting an equipment foreman or manager to an issue much earlier. Additionally, GPS guidance systems can and are being used with earthmoving equipment to cut, grade, and contour project site landscapes.

Another technology that has emerged prominently at the job site is digital photography, both still frame and video. These photographs can be large in size, prolific, and stored in various places, from web-based storage containers to local drives on laptops and tablets to network drives at the office to the "cloud."

Scheduling software has been around for many years. Utilities and builders of power-generation facilities are paying more attention to schedules these days, not just for help in planning and sequencing the work but also as an aid to help document or defend delay or cost overrun claims when necessary. The cost to rebuild schedules of what actually happened, after the fact, in the case of a claim and/or lawsuit, can be staggering, so maintaining every revision is important. Scheduling software is designed to do this.

How to Select

Before heading out to the marketplace to acquire new systems, it is essential to take a step back to identify the requirements from the new system and how the proposed solution will interact with others already in place. This means stepping outside the current business processes and asking questions about what is needed and how tasks are accomplished. Sufficient time must be spent documenting requirements objectively without regard for one product or another, or consideration of how the process had been accomplished in the past. These requirements will establish the baseline against which to measure various solutions.

A common practice when evaluating a new solution is a software demonstration by the vendor of that system. Unfortunately, demonstrations are often rushed and become a short-duration experience of seeing what the vendor wants to show off rather than addressing the specific need.

The demonstration should be of sufficient time to give the team a chance to see the product in action and evaluate important functions, as well as to understand the vendor company and its implementation approach. It is not advisable to view more than two or three products when selecting a new product. Typically, after three product demonstrations, the product differences begin to blend, and it becomes a time-consuming and confusing process to separate these differences afterwards.

But ultimately, it is not just about the product. The team should also verify that the vendor company is in good financial position. The vendor company needs to be financially successful in order to support the product and continue to develop it. It should have a reputation for taking care of its customers and assisting them during the implementation process. It is also important to call a few vendor-provided references to learn more about how the product performs and vendor behaves after the sale.

Deployment Models

As with many developments in the fast-moving field of technology, cloud computing is becoming very popular in the technology industry. Readers of this chapter some years from now may wonder why there was any reluctance to adopt this approach. Today, it is not yet the predominant platform, although the arguments in favor of this technology are well-founded.

Systems have gone through their share of deployment models. During the 1980s, when organizations could not afford their own midrange computer system, vendors would develop software and provide access to that software system via landline connections. It was the first application service provider (ASP). At the time, one did not care so much where the computer or data was, just that the system was operational each day. As personal computers evolved and became more affordable, software was written to run on corporate servers and personal computers rather than within a stand-alone data center. But as the size of the computing environment and complexity to run it increased, managers rethought the necessity of owning and maintaining all of that computing infrastructure. And with the increased accessibility and dependency provided by the Internet, it became easier to provide access to the software and house a company's data in an arguably safer environment in a modern data center.

Software as a service (SaaS) and ASP are precursors to today's cloud environment. The premise is largely the same, although there are some technical differences. Among the key differences are who owns the rights to the software and how the software is paid for. In a traditional software license agreement, a company would pay for the full right to use the software, typically according to the number of users. That software license would then be conveyed to the company for its use. This method has advantages but also carries with it a degree of risk. If the using company shrinks or decides the system no longer meets its needs, the company has already prepaid for the software in full, for naught. The cloud or SaaS model is more closely a pay-as-you-go model, where a certain fee is established each month for the use of the software (like cable TV). This eliminates the large up-front payment and means that the company can scale the user count up or down depending on the need. This model is also very similar to the automobile leasing idea. Over time, if you lease the car for more than five years, you will pay more than buying it outright. However, if you are not sure how long you will want the car or if it is the right model for you, the lease can be practical. Whichever model is chosen, performance of a new system in the field is critical. Many well-intended and well-chosen solutions have failed to meet expectations simply because the system ran too slowly from the field office.

Good connectivity remains an issue at some job sites even today. However, most sites can get some level of connectivity and then simply use good servers and network management tools to ensure adequate performance from the field. Many companies are now using the fourth generation (4G) of mobile phone communication technology standards and services for situations where a consistent wireless local area network, more commonly known as a WiFi connection, is not available. Finally, if connectivity just is not available, some applications or functions within applications have been designed to run in a detached mode with upload and syncing functions available when connection is re-established.

How to Implement

Ownership of software or hardware is not what provides the value; adoption and use are. For these reasons, it is critical that the implementation is managed carefully and with an eye toward the end users. There are several important phases to implementation. No matter what system or application is being implemented, the steps remain roughly the same, though durations can change. A brief description of the primary steps follows:

- Planning—It is said that failing to plan is planning to fail. While that expression has been around for a long time, it could not be more appropriate when it comes to software and technological tool implementation. A good implementation plan for using modern technologies to manage construction projects is just as important as a good project execution plan for the actual construction of a power plant or renovating an existing one. A plan should take into consideration the team, their roles and responsibilities, schedule, risks, and objectives. This consideration should be clearly communicated to everyone involved. As in all other facets of construction management, a good plan also provides for communication on a regular basis, how to handle issues, and what the appropriate scope is. Changing scope during the implementation of technology to enhance construction management can be just as costly as changing the design in a construction project, once underway.

- Design—This is a very important step in the process, in which the company and the vendor or consultant work together to make decisions on how the system is to work and what the expectations are.

- Setup—This function is among the most technical during the implementation steps and should only be done after the design is complete. Often times, various settings have to be established in a new system based on how the organization wants it to perform. This step can also include security consideration, integration requirements, and changes to existing reports and screens.

- Testing—This is, in many ways, the most critical function. Most software and technology systems have many options and settings that control the way the systems actually work. Improperly set, the systems may not work at all, or will not work as expected. Testing typically begins once the process has been designed and the software is set up. For larger systems, a testing script is recommended that includes the transactions to test, expectations, room for results, and follow-up actions as required. A company should not proceed with converting or going live on a new software or technological system until sufficient testing has been completed. This should include integrations with other systems as well. Insufficient testing by the implementation team means that the testing function informally falls to the end users, and that is the wrong group to be debugging a new process.

- Documentation—Many of the software and technology systems come with documentation but rarely come with something closer to what a user really needs, a customized "cheat sheet" if you will. The system documentation is seldom tailored to the unique procedures or processes of the organization. Developing user documentation does not have to be time-consuming, but it is very important. It can be used in conjunction with end-user training and left behind as a guide as people start using the system. Anything and everything that helps users adapt to the new system and procedure is generally good. Today, some companies are setting up their own wiki sites to house user documentation, or they may have a library maintained in a common storage location, such as Microsoft's SharePoint or other content management system.

- Training—There are two forms of training during implementation. One is called system training, and the other is end-user training. System training is provided to a few; end-user training is provided to everyone who will use the application. System training is necessary for those responsible for setting up the system, making decisions about how it works, and testing the application. This level of training is provided by the software vendor or an independent consultant. End-user training happens closer to the date when the software is ready to be implemented. End-user training only covers what the user typically needs in order to begin using the software correctly. It can be provided by the vendor, but is often better provided by the implementation team members for the organization that will be using the software and technology.

New technology and solutions are becoming not only prevalent but also necessary tools. There are a myriad of choices for most of the functions performed in the field. Having a structured selection process in place to ensure a good decision is a critical first step to ensure an appropriate choice. Following that, an effective and well-thought-out implementation process will ensure that whatever the solution is, it will be well received and meet expectations. Deploying solutions in the field is no longer a luxury. It is a necessity.

The oft-spoken desire of using modern technology to manage construction projects has been to improve coordination of information on drawings between various disciplines such as civil, steel, mechanical, and electrical; streamline estimating; build from design models; and minimize cost overruns. In addition, it is often said that much of the drain in construction productivity is due to a lack of timely information. However, there are some caveats. According to a 2009 *Engineering News-Record* article, "Expecting a Win by Taking on BIM," the most effective approach is to "get everyone in one room where they all (have) laptops. We tried overhead projectors, smart boards and all that stuff, and what we found most effective was having everybody in the same room."[1] But times have changed, and maybe the better tool today is a tablet, something not very common in 2009. It has been said that tablets are (today's) new tool in the construction industry's tool belt. What will there be in the year 2019, 2029, or 2039?

Some Examples

The intent of this chapter is to describe the new tools, methods, and technologies that the people in the field are starting to use on a daily basis, so here is a list of some that we have gleaned from various sources, mostly with a view toward touching real-time information, only once, and resulting in immediate actions:

- Photo-management tools that turn job-site photos into mineable data, searchable in a content-management system.
- Tagging photos to drawings and specs.
- Tagging construction plans (and photos) by trade.
- Tools to incorporate 3D, 4D, and 5D point-cloud technology into BIM. Essentially, this is a way of taking laser scans and photos with handheld devices and creating intelligent models.
- Layering the photos and scans so they can be manipulated similar to using the older practice of tracing paper.

- Modeling existing equipment, structures, pipes, and conduits.
- Laser scanning tools replacing traditional surveying tools.
- BIM modeling using quick response (QR) codes to anchor location.
- QR code stickers linking to prior data and laser scans (e.g., a wall's content just before closing or cladding—records of as-builts of now-hidden conditions).
- Bar codes for identifying materials and equipment, as well as personnel.
- Scanners for determining levelness and for preparing topographic maps.
- 3D printing, especially for modeling purposes (such as complex lifts).
- Hyperlinking electronic documents to each other, such as plans and specs.
- Animated work packages in sync with engineering tasks.
- Calculating work put in place by overlaying job-site photos on a 3D model.
- 4D (time) and 5D (money) look-ahead simulations.
- Printing layout lines using laser-equipped robots.
- Automatic (and even remote-controlled) welding.
- Telematic (via satellite uploads) communications for use in managing heavy equipment—hours of use, diagnostics in the event of failure, and theft protection.
- Selling surplus project products online.
- Streamlined invoicing preparation, transmittal, and approvals.
- VoIP (Voice over Internet Protocol).
- Internet searches on almost any topic.

And finally, now there are thousands of apps (applications) for tablets, smart phones, and other mobile telecommunication devices (often replacing the manual clipboard) that can do the following:

- Access job-site cameras.
- Calculate cost estimates with preloaded prices for labor and materials.

- Perform trade-specific construction calculations such as angles, concrete (sand/cement ratios) quantities, quantity of brick-mortar ties, elevation slope requirements, and the like.
- Use the GPS function of a smart phone to deliver satellite images of job sites and existing structures on those sites.
- View CAD files.
- Access *all* project drawings from mobile devices.
- Automatic cross-linking of PDF drawing updates to all users' mobile devices.
- Render 3D drawings for commissioning, punch-listing, and documenting.
- Create forms for project management (time sheets, quality checklists, RFIs, transmittals, job reporting, due and pending work sheets, etc.).
- Track project documentation.
- Track project progress in 3D.
- Instant messaging (IM).
- Capture notes and voice memos and synchronize them across all devices.
- Convert voice memos to print media for electronic transmission.
- Take photos and geo-tag them into blueprints and plans—especially useful for work that will be enclosed and no longer visible.
- Distance measurements.
- Acoustic measurement and management
- Heat index measurement (for OSHA compliance).
- Crane hand signal depiction.
- Rigging load capacities of shackles and slings.
- Crisis management—the app knows where the incident occurred, the app user's job role, and whom to contact to launch the cascade of responses often required to control the rumor mill and appease others.
- "Man down" fall notification.

- "Lone worker" button to request periodic check-in until alert is canceled.
- Punch list with pictures, notes, etc.
- (Extra) work authorizations.
- Job tool tracking and control.

Other Important Bits of Information

Interestingly, much of the technology that is behind many of the above tools and apps was initially developed for gaming use, both for home use and casino-type machines. The jury is still out as to the actual value of using these methods, technologies, and processes, and as this continues to develop, the return on the investment will continue to be evaluated.

There are stability issues. For example, how does one keep many of these "powered" tools powered? Just now arriving, there are tools such as charging bowls where one can place several devices into a bowl-like device that uses magnetic resonance to charge them. Other devices, some of them solar powered, are designed to accept USB connections for use in charging tools. And others are semiconventional electrical connecting devices that use magnets to "suck" the tool's charging cord into a plug.

Also, what happens when computer servers go down, whether due to internal or external reasons? For example, if electric power is interrupted and a server can no longer transmit data, the user will not have access to information stored in the cloud. A backup plan must be part of any IT procedure.

There also may be periodic problems of phone coverage. In some areas of the United States and especially in some parts of the world, phone coverage is spotty or even nonexistent. Although these issues can be resolved, their resolution is not always cost-effective. For example, a contractor planning a short turn-around at a remote power plant could be in such a situation.

And there are support issues. What about support in the field when these technologically driven tools do not work? One does not want the time saved using the technology to be spent by the time required to support it. Many organizations have small IT departments, and they cannot be everywhere at once for training and repairing. Finally, it is still a fact of life that many construction field personnel are too embarrassed to admit that they do not know how to use this new technology or ask for help. Younger workers are excited to use these tools while some of the older workers are reluctant, and this creates a dichotomy.

An actual case in point: A younger field superintendent and one with many years of experience were paired up to run a boiler repair project. They shared one end of a construction site trailer, using a 4×8-foot plywood board supported on filing cabinets as a desk top. The right end, where the younger superintendent worked, was very neat, with only a laptop computer, a smart phone, and a few forms. The left end, where the experienced superintendent sat, was littered with small-scale drawings, notebooks, pens and pencils, a calculator, and a flip phone. They put a sign above this workstation that read from left to right: "**Analog Age … Digital Age.**" The project was completed on time and within budget, which is to say that no single approach—digital or the older analog—is always the right approach, and sometimes a blend of the two is even better.

Another concern is real-time access to this technology for those in the field who are expected to use and implement it. There are statistics that suggest that less than half of the intended users have the tools or training to use this technology. But if they do, the advantages can be enormous, including automatically synchronizing project plans, specs, and photos, along with marked-up notes and attachments, which can then be sent to everyone's mobile devices. Working together this way is also a surefire method of ensuring that everyone is using the same, latest version of data available.

Finally, there is the issue of storage medium. The ubiquitous cloud storage seems an ideal answer, except that it is still fairly new and therefore unproven regarding stability and security. In the same vein, there is the joint sharing of data and information across company borders. How safe and secure is your information? This leads into the next topic: security.

Security

Security in cyberspace is a nebulous issue. We all want data interoperability, and we want the ability to collaborate. But then, we also want to keep our data secure and our "doors" locked. Unfortunately, this is not completely achievable; but that does not mean we cannot be smart and protect ourselves from most cyber threats. While many corporate IT departments have very stringent protocols in place to accomplish this, these protocols often result in restrictions to the very activities many want to use with this new technology.

So how can this balance be achieved? Slowly, but surely, industry groups are working to get users together, often in one- and two-day conference sessions, to share needs and successes. The take-away from these conferences can be a more in-depth understanding of how to specifically address security issues in one's own organization.

In February 2013, President Obama issued Executive Order 13636, titled "Improving Critical Infrastructure Cybersecurity." This order instructed the U.S. National Institute of Standards and Technology (NIST) to develop a voluntary cybersecurity framework that would provide a "prioritized, flexible, repeatable, performance-based, and cost effective approach for assisting organizations responsible for critical infrastructure services to manage cybersecurity risk." Although this order is pointed at large installations, the spin-off from the development of this cybersecurity network will be of immense value to all who are charged with protecting their own company software and technology.

But ultimately, it is also the individual user's responsibility to protect the company and its access, similar to locking the doors of the office building before going home at night and/or using a security service to keep an eye out for the unscrupulous. That is to say, make sure passwords are strong and changed frequently. Make sure antivirus programs are installed, used, and kept updated. Do not click on software links that lead into the unknown. And never, ever give out privileged information to anyone through cyberspace without using proper protections and protocols. In this way, some of the risks of security breaches can be mitigated, and we can all sleep easier at night.

Keeping Abreast of the Newest

Ultimately, we would like to leave the reader with a method of being able to stay on top of new developments. In other words, we would like to leave a path the reader could follow to discover new tools that are not yet available, and once they do become available, a way to know about them—a sort of timeless ending.

Unfortunately, there is no central source for this. Technology systems, software, and other not-yet-invented concepts and tools are always emerging. Sometimes, they come out of necessity; other times, out of curiosity and/or experimentation. The unfortunate situation is that there is no central center of excellence that collects all of this. One can attend various industry conferences, read some of the numerous industry publications, or surf the Internet and discover what many or our peers are doing. Also, don't forget internal IT departments. The people staffing these departments often know about new technology and tools, but may not always realize that these new systems could find a home on the construction site.

The bottom line is that technology and mobile field devices are no longer optional considerations. To maintain business agility and remain competitive, they must form an integral part of any project execution plan. One must be able to be at the head of the class, so to speak, and know what technologies are coming even before they arrive. One must spend the time and effort to get plugged in using some of the above methods.

Summary

Wow! Not sure what we just went through. A treatise on something many of us are still trying to wrap our minds around. Of course, there are some who get it right away—technology is the driving force of humanity and has been since the industrial revolution. But there are others of us who are so comfortable in our own zone that the suggestion of change is scary. What do we do? Deviating from the format of the previous chapters, this is really not a summary of this chapter. It stands on its own. We (the author and contributors) would appreciate feedback to enhance this subject matter. Please contact us at info@ConstrBiz.com.

ENJOY!

Reference

1 Sawyer, Tom. "Expecting a Win by Taking on BIM." *Engineering News Record*, May 4, 2009, p. 35.

Owner's Construction Estimate Checklist

Yes	No	Specification overview	Comments
☐	☐	Scope clearly identified?	_____
☐	☐	Equipment and materials specified?	_____
☐	☐	Schedule limitations known?	_____
☐	☐	Constructibility issues reviewed?	_____
☐	☐	Bonus/penalties considered?	_____

Yes	No	Resource review	Comments
☐	☐	Adequate time to prepare estimate?	_____
☐	☐	Adequate resources to prepare estimate?	_____
☐	☐	List of assumptions	_____

Yes	No	Project conditions	Comments
☐	☐	Labor availability considered?	_____
☐	☐	Station/contractor scope split defined?	_____
☐	☐	Additional work scope possible?	_____
☐	☐	Safety history considered?	_____
☐	☐	Nurses/EMT priced?	_____
☐	☐	Mobilization/demobilization included?	_____
☐	☐	Specialized rigging priced?	_____
☐	☐	Specialized machining priced?	_____
☐	☐	Start-up support included?	_____
☐	☐	Contractor interfacing investigated?	_____
☐	☐	Site security issues considered?	_____
☐	☐	Temporary power included?	_____
☐	☐	Trash removal included?	_____
☐	☐	Sanitary facilities included?	_____
☐	☐	Insulation, lagging, and refractory included?	_____
☐	☐	Asbestos removal/disturbance considered?	_____
☐	☐	Lead paint considered?	_____

☐ ☐ Soil contaminants considered? _____
☐ ☐ NDE included? _____
☐ ☐ Stress relieving included? _____
☐ ☐ Self-insurance planned? _____
☐ ☐ Labor and supervision cost of living included? _____
☐ ☐ Taxes included? _____
☐ ☐ Other items? _____

Yes No Sanity check   **Comments**
☐ ☐ Total man-hours realistic? _____
☐ ☐ Dollars per man-hour realistic? _____
☐ ☐ Total dollars realistic? _____

Contractor's Construction Estimate Checklist

Yes	No	Specification overview	Comments
☐	☐	Scope clearly identified?	_____
☐	☐	Equipment and materials specified?	_____
☐	☐	Schedule limitations known?	_____
☐	☐	Liquidated damages understood?	_____
☐	☐	Bonus/penalties considered?	_____
☐	☐	Constructibility review performed?	_____

Yes	No	Resource review	Comments
☐	☐	Adequate time to prepare estimate?	_____
☐	☐	Adequate resources to prepare estimate?	_____
☐	☐	Proposal number issued? (implies approval to bid)	_____

Yes	No	Specification review requested/performed	Comments
☐	☐	Quality assurance/NDE	_____
☐	☐	Commercial/legal	_____
☐	☐	Construction engineering	_____
☐	☐	Rigging engineering	_____
☐	☐	Welding engineering	_____
☐	☐	Labor relations/safety	_____
☐	☐	Accounting (cash flow, taxes, D&B)	_____

Yes	No	Estimate/proposal development	Comments
☐	☐	Mobilization/demobilization included?	_____
☐	☐	Specialized rigging included?	_____
☐	☐	Specialized machining included?	_____
☐	☐	QA specialist included?	_____
☐	☐	NDE included?	_____
☐	☐	Stress relieving included?	_____
☐	☐	Authorized inspector included?	_____

☐ ☐ Safety officer included? _____

☐ ☐ Site security included? _____

☐ ☐ Temporary power included? _____

☐ ☐ Trash removal included? _____

☐ ☐ Sanitary facilities included? _____

☐ ☐ Setting, insulation, and lagging _____
(SIL) (off/on) included? _____

☐ ☐ Asbestos removal/disturbance considered? _____

☐ ☐ Lead paint considered? _____

☐ ☐ Start-up support included? _____

☐ ☐ Payroll/accounting support included? _____

☐ ☐ Sales and other local taxes included? _____

☐ ☐ International taxes and tax prep fees included? _____

☐ ☐ Other costs included? _____

Yes No   **Other considerations**   **Comments**

☐ ☐ Labor availability _____

☐ ☐ Customer/partner scope split defined? _____

☐ ☐ Additional work possible? Amount? _____

☐ ☐ List of assumptions included? _____

☐ ☐ Interfacing with other contractors? _____

☐ ☐ Physical site conditions investigated? _____

☐ ☐ Subcontractor pricing obtained (in writing)? _____

Yes No   **Sanity check**   **Comments**

☐ ☐ Total man-hours and total dollars realistic? _____

☐ ☐ Dollars per man-hour realistic? _____

☐ ☐ Total dollars realistic? _____

Contract Responsibilities Matrix

Item	By Owner	By Contractor	By Sub
1. CIF	_____	_____	_____
2. Harbor charges: (dockage fees, unloading fees, demurrage charges, permits, licenses or other levies)	_____	_____	_____
3. Custom or import duties	_____	_____	_____
4. Transport of materials and unloading (port to job site)	_____	_____	_____
5. Unloading of materials at job site	_____	_____	_____
6. Transport of material at job site	_____	_____	_____
7. Material laydown and storage area prepared for heavy equipment movement (state acres required)	_____	_____	_____
8. Maintenance of storage area	_____	_____	_____
9. Inside storage, warehouse (state-required square footage)	_____	_____	_____
10. Job-site office & facilities (state-required square footage)	_____	_____	_____
11. Furniture, equipment, and air conditioning for office	_____	_____	_____
12. Job-site tool room	_____	_____	_____
13. Job-site change room	_____	_____	_____
14. Housing for contractor personnel	_____	_____	_____
15. Housing for vendor representatives	_____	_____	_____
16. Housing for locally hired personnel	_____	_____	_____

17. Housing and mess facilities for workers   _____ _____ _____

18. Medical facilities   _____ _____ _____

19. Transportation for contractor personnel   _____ _____ _____

20. Transportation for vendor representatives _____ _____ _____

21. Transportation for workers   _____ _____ _____

22. Sanitary facilities
    Administration
    Staff   _____ _____ _____
    Workers   _____ _____ _____

23. Water
    Construction
    Drinking   _____ _____ _____

24. Security
    Job perimeter
    Job work area   _____ _____ _____

25. Fire protection
    Job site
    Work area   _____ _____ _____

26. Trash and debris removal   _____ _____ _____
    (if offsite, state distance)

27. Safety
    Job site
    Work area   _____ _____ _____

28. Electric power (state-required KVA)   _____ _____ _____

29. Distribution of electric power   _____ _____ _____

30. Communications
    Telephone
    Telefax   _____ _____ _____
    Radio   _____ _____ _____
    Computer link/satellite   _____ _____ _____

31. Air for construction use   _____ _____ _____
    (state-required CFM)

32. Distribution of air   _____ _____ _____

33. NDE ____ ____ ____

34. Interpretation of NDE ____ ____ ____

35. Stress-relieving equipment ____ ____ ____

36. Welder qualifications
    Test coupons ____ ____ ____
    Cost of qualifications ____ ____ ____
    Cost of welder training (if required) ____ ____ ____

37. Interpretation of welder qualification ____ ____ ____

38. Weld rod
    Pressure parts ____ ____ ____
    Nonpressure parts ____ ____ ____

39. Hydrostatic test
    Water ____ ____ ____
    Chemicals ____ ____ ____
    Test pump and gauges ____ ____ ____
    Fill pump ____ ____ ____
    Labor ____ ____ ____

40. Chemical cleaning
    Boil out
    • Chemicals ____ ____ ____
    • Operation ____ ____ ____
    • Assist Labor ____ ____ ____
    Acid cleaning
    • Chemicals ____ ____ ____
    • Operation ____ ____ ____
    • Assist labor ____ ____ ____
    • Disposal of chemicals ____ ____ ____

41. Grout
    Materials ____ ____ ____
    Labor ____ ____ ____

42. Fit-up bolts (temporary bolts usually ____ ____ ____
    not supplied with material)

43. Paint
    Final paint
    • Material   _____ _____ _____
    • Labor   _____ _____ _____
    Touch-up paint
    • Material   _____ _____ _____
    • Labor   _____ _____ _____

44. Scaffolding
    For base project   _____ _____ _____
    For inspections   _____ _____ _____

45. Weather protection
    For stored materials   _____ _____ _____
    For work area   _____ _____ _____
    For construction equipment   _____ _____ _____

46. Local permits, licenses, fees, or other levies   _____ _____ _____

47. Insurance
    Builders' all risk   _____ _____ _____
    Property damage   _____ _____ _____
    Public/third-party liability   _____ _____ _____
    Vehicle insurance   _____ _____ _____

48. Temporary facilities and consumables   _____ _____ _____
    required for construction

49. Interpreters   _____ _____ _____

50. Air-pressure test
    Temporary blanks   _____ _____ _____
    Fan or blower   _____ _____ _____
    Labor to conduct test   _____ _____ _____

51. Refractory, insulation, lagging
    Materials   _____ _____ _____
    Labor   _____ _____ _____
    Special tools/fab facilities   _____ _____ _____

52. Performance bond   _____ _____ _____

53. Payment bond   _____ _____ _____

54. Completion penalty fee (L/Ds), amount   _____ _____ _____

Job-Site Visit Information Sheet

Name of prospect _____ Date of visit _____

Work description _____

Address of prospect _____

Person making visit _____

Name and position of person _____
   representing customer   _____

Consulting engineer _____

Specifications available   Yes ☐   No ☐
   Date erection starts (estimate) _____

Job-site location

_____
Route or Street   Town

_____
County   State

Site plan available from customer?   Yes ☐   No ☐

Describe existing buildings: _____

_____

Access and Storage Areas

Access Road

Describe length, type, surface, _____
   width, sharp curves.   _____

Any change necessary? _____

Who maintains? _____

Parking Area

Describe distance from work _____
area and change room.   _____

Buses required?   Yes ☐   No ☐
Who provides? _____

Delivery Point

What is location of nearest common   _____
carrier free delivery point? Any   _____
improvements needed or expected?   _____
Are other contractors using?   _____

Storage Area

Describe area, including access, soil   _____
conditions, and overhead lines.   _____
Share with anyone? Describe   _____
instrumentation storage. How to   _____
protect against damage and pilferage? _____

Access from: Three sides ☐   Two sides ☐   One side ☐

Will temporary buildings be required?   Yes ☐   No ☐
If yes, what kind? _____
Who supplies? _____

Will electricity be required?   Yes ☐   No ☐
Who supplies? _____

What is available?   Voltage _____   Cycle _____
Power _____
If lines to be run, how far? _____

Will telephone be required?   Yes ☐   No ☐
Who supplies? _____
If lines to be run, how far? _____

Subassembly Area

Describe. Will it be in storage area   _____
or at work site? Is it shared?   _____

Distance from common carrier free delivery point to storage area. _____

Method of material movement from _____
    delivery point to storage area. _____
    Any interferences? _____

Distance from storage area to work area. _____

Method of movement of material from _____
    storage area to work area. Any _____
    interferences? Describe all access _____
    to work area. Can subassemblies _____
    be handled? Will other contractors _____
    cause interferences? _____

Facilities

Field Office

Location, size, type required. _____

Required electric power. _____

Any present facilities. _____

Who supplies? _____

Tool Room

Location. _____

Dark room location. _____

Who supplies? _____

Sanitary Facilities

Who furnishes? _____

Location, number, and type? _____

Can there be urinals in building? _____

Change Rooms

Type, size, number needed. _____

Electric power needed? _____

Type of heat? _____

Distance from work area? _____

Who supplies? _____

First Aid

Who furnishes? _____

Distance to nearest doctor and hospital? _____

Vehicles required for transportation to doctor or hospital?   Yes ☐   No ☐

Electric Power

What is presently available?   Voltage _____   Cycle _____
  Power (KW or KVA) _____ What is present location? _____

What will have to be done to bring it in? _____
  Are transformers required?   Yes ☐   No ☐
  How will power be made available for _____
  welding machines and other equipment? _____

Lighting

What exists in the way of temporary lighting? _____

Who will furnish? _____

Where is power available for lighting? _____

What KVA is required? _____

Water

Is construction water available?   Yes ☐   No ☐
  If so, where? _____

Is drinking water available?   Yes ☐   No ☐
  If so, where? _____

Is ice available?   Yes ☐   No ☐
  If so, where? _____

Can electric water coolers be used?   Yes ☐   No ☐

Compressed Air

Who will furnish? _____

Is operator required? Yes ☐ No ☐

Needed cubic feet per minute _____
    Length of time required _____
    Will manifolding be required? Yes ☐ No ☐

Cranes

Number needed _____ Capacity _____ Type _____
Number needed _____ Capacity _____ Type _____
Number needed _____ Capacity _____ Type _____
What is the local rental situation? _____

Elevators

Can present ones be used? _____
If temporary hoists are needed, what type? _____
Where installed? _____
Heating and ventilation for trades on job: _____
    What will be required?   _____
Debris and scrap handling facilities: _____
    What will be required?   _____
    Who will furnish?   _____

Personnel

Proposed number and type of superintendents:

No.	Type	Timing
___	_____	_____
___	_____	_____
___	_____	_____
___	_____	_____
___	_____	_____
___	_____	_____
___	_____	_____

Name of lead superintendent: _____

Unions: Locals having jurisdiction.

Craft	Lodge No.	Location	Distance to job

Workload in area _____

_____

_____

_____

Hours being worked in area _____

Estimate of availability of qualified people _____

Other

Any unusual items in specifications _____

_____

_____

_____

Remarks _____

_____

_____

_____

_____

_____

Signature of person making visit

Business Controls Checklist

Yes No Project Management

☐ ☐ Project execution outline
☐ ☐ Business practices review
☐ ☐ Extra work authorizations
☐ ☐ Contract on-site
☐ ☐ Estimate on-site
☐ ☐ Schedule on-site
☐ ☐ Contract abstract on-site
☐ ☐ Cash flowchart on-site
☐ ☐ Scope clearly identified
☐ ☐ Liquidated damages understood
☐ ☐ Bonus understood
☐ ☐ Delay claims notification understood
☐ ☐ Drawings on-site
☐ ☐ Bills of materials on-site
☐ ☐ Weekly financial report
☐ ☐ Compare weekly financial report with home office cost reports.
☐ ☐ Special bonuses accrued
☐ ☐ Daily log
☐ ☐ Weekly status report
☐ ☐ Weekly progress graph
☐ ☐ Weekly labor progress work sheets
☐ ☐ Extra work rates with client in file
☐ ☐ Extra work rates with subcontractors in file
☐ ☐ Minutes of weekly customer meeting
☐ ☐ Notification to customer of delays, problems, and/or potential claims
☐ ☐ Plan for storing tools and equipment after job
☐ ☐ System for tools and consumables control
☐ ☐ Storage system for expensive items

☐ ☐ Tool inventory
☐ ☐ Brass/time card system
☐ ☐ Daytime security guard
☐ ☐ Night/weekend security guard
☐ ☐ Lunch box check
☐ ☐ Material receipt and inspection reports
☐ ☐ Welding forms on-site
Notes: _____

Yes No Administrative

☐ ☐ Craftsmen job applications
☐ ☐ Employment record forms available
☐ ☐ W-4
☐ ☐ I-9
☐ ☐ Local labor employment contracts available
☐ ☐ Union agreements available
☐ ☐ Visas
☐ ☐ Work permits
☐ ☐ Verification that payroll deductions are current
☐ ☐ Job time report
☐ ☐ Time distribution report
☐ ☐ Authorized inspector purchase order (PO)
☐ ☐ Job-site record control log
☐ ☐ Job-site close out records transmittal
☐ ☐ Petty cash accounting
☐ ☐ Field expense report
☐ ☐ Lay-off notices in file
☐ ☐ Insurance coverage
☐ ☐ Payroll data backed up
☐ ☐ Hold harmless statement for renting to third parties
☐ ☐ Backcharge claims
☐ ☐ Extra work authorization
☐ ☐ Time, material, and rental records
☐ ☐ Bank account reconciled
☐ ☐ Billing procedures available
☐ ☐ Job-site address available to vendors
☐ ☐ E-mail set up
☐ ☐ Company rules given to employees

☐ ☐ Notice of workers' comp posted
☐ ☐ First aid facilities
☐ ☐ Ambulance/hospital facilities and plans

Yes No Purchasing

☐ ☐ Three quotes in file
☐ ☐ Single source letter in file
☐ ☐ Field purchase orders in file
☐ ☐ Purchase order requisitions
☐ ☐ Purchase order in file
☐ ☐ PO supplements in file
☐ ☐ Vendor backcharge log
☐ ☐ Subcontractor extra work order log
☐ ☐ Contractor insurance certificates available
☐ ☐ Gasoline tax exemption certificate
☐ ☐ Sales tax exemption certificate

Yes No Labor

☐ ☐ EEO poster displayed
☐ ☐ Workers' comp certificate displayed
☐ ☐ Sexual discrimination poster displayed
☐ ☐ Sexual harassment notice displayed
☐ ☐ Antidrug policy poster displayed

Yes No Safety

☐ ☐ Safety manual on-site
☐ ☐ Pretask plan available
☐ ☐ Job safety plan available
☐ ☐ MSDSs in file
☐ ☐ Hazard communication program available
☐ ☐ Confined space program available
☐ ☐ OSHA posters displayed
☐ ☐ Scaffold tags used
☐ ☐ Safety glasses used
☐ ☐ Hard hats used
☐ ☐ OSHA 300 log displayed
☐ ☐ Competent person identified
☐ ☐ Lock-out/tag-out procedure used

☐ ☐ Employee safety handbooks issued
☐ ☐ Employee safety handbook receipt in file
☐ ☐ Tool box safety meetings
☐ ☐ Work rules posted
☐ ☐ Awards program
☐ ☐ Local medical providers arranged
☐ ☐ Graph of safety indicators
☐ ☐ Safety team in place
☐ ☐ Coffee machines must be commercial grade
☐ ☐ Self-inspection records available
☐ ☐ Formal process for employees to report near misses
☐ ☐ Near misses being tracked
☐ ☐ Outside service to clean up blood
☐ ☐ Primary responder for each 50 workers
☐ ☐ Any brown, two-prong extension cords?
☐ ☐ Plan in place to handle the OSHA inspector
☐ ☐ "Mock" OSHA inspection made
☐ ☐ OSHA inspection kit available
☐ ☐ Forklift training—site specific
☐ ☐ Record keeping per requirements

Yes No Quality

☐ ☐ Nonconformance reports
☐ ☐ NCR Log
☐ ☐ Welders' qualification records in file
☐ ☐ Quality control plan available
☐ ☐ Authorized inspector involved
☐ ☐ QA manual on-site
☐ ☐ NDE plan available
☐ ☐ NDE subcontractor identified

Courtesy of Construction Business Associates, LLC.

Index

D

M

webcams, 338
weekly progress report, 324–325, 334
weighted values, 83–85
welders, 220
welding, 69, 217, 266–267
    cost impact of, 227–228
    productivity measurement of, 227–
        228, 289–293
    progress graphs, 290–293, 321–
        323
    training, 129
WiFi, 344
work breakdown structures (WBSs),
    35–36, 111, 265
    assigning value to, 273–275
    earned value calculations and,
        273–274
    estimator review of, 90, 105–106
    for financial status reports, 268
    man-hours earned in, 274–275
    productivity and, 273–275, 289
    purposes of, 273
    scheduling, 275–276
workers' compensation insurance, 12,
    148, 158, 250
    fraudulent claims for, 258–260
    owners' provision of, 106, 243,
        250–251
    requirements of, 191–192
    safety and, 97, 234–235
    supervisory process for, 258–259
work-in-progress activities, 245–247
wrap-up policies, insurance, 192
written safety program, 237–238

Z

zero injury concept
        industry averages and, 255–256
        injury-hiding and, 258
        obstacles to, 256
        requirements of, 256
        site support needs for, 257